NUCLEAR MAGNETIC RESONANCE

NUCLEAR MAGNETIC RESONANCE

BY

E. R. ANDREW
PH.D., F.R.S.E.

Professor of Physics in the University of Wales
(University College of North Wales, Bangor)

CAMBRIDGE
AT THE UNIVERSITY PRESS
1958

PUBLISHED BY
THE SYNDICS OF THE CAMBRIDGE UNIVERSITY PRESS

Bentley House, 200 Euston Road, London, N.W. 1
American Branch: 32 East 57th Street, New York 22, N.Y.

First Edition 1955
Reprinted 1956
1958

First printed in Great Britain by Robert MacLehose and Co. Ltd
The University Press, Glasgow
Reprinted by offset-Litho by
Jarrold and Sons Ltd., Norwich

TO MY WIFE

AUTHOR'S PREFACE

Since the first successful experiments on nuclear magnetic resonance were carried out at the end of 1945 there has been a remarkable expansion of this new field of research, with applications reaching into many branches of physics and into other sciences as well. The essential features of the phenomenon are now fairly well established, and the experimental techniques are well developed. This monograph attempts a first comprehensive review of the subject.

I owe a great debt of gratitude to Professor E. M. Purcell, in whose laboratory at Harvard University in 1948–9 I first worked in the field of nuclear magnetic resonance. I am also greatly indebted to Professor J. F. Allen, who suggested this line of work to me and encouraged me to continue it at St Andrews University. A special word of thanks is due to Dr D. ter Haar, Professor E. M. Purcell and Dr D. Shoenberg, all of whom read the entire manuscript before it was sent to the publishers, and whose many suggestions have greatly improved the book. I have derived great benefit from the shorter reviews and classic articles listed in the first page of the bibliography; the lucid articles of Professor G. E. Pake (1950) were particularly illuminating.

I wish to thank the editors of the *American Journal of Physics*, the *Journal of Chemical Physics*, *Physica*, the *Physical Review*, the *Proceedings of the Royal Society*, *Progress in Nuclear Physics*, and the *Transactions of the Faraday Society* for permission to reproduce a number of diagrams, and I wish to thank the authors of the articles from which the diagrams were taken.

E. R. ANDREW

August 1954

CONTENTS

CHAPTER 5

Nuclear Magnetic Resonance in Liquids and Gases

CHAPTER 6

Nuclear Magnetic Resonance in Non-metallic Solids

CHAPTER 7

Nuclear Magnetic Resonance in Metals

CHAPTER 8

Quadrupole Effects

8.1. Introduction, p. 203. **8.2.** Quadrupole effects in liquids and gases, p. 204. **8.3.** Quadrupole effects on the nuclear magnetic resonance spectrum for solids, p. 206. **8.4.** Quadrupole relaxation effects in solids, p. 211. **8.5.** Pure quadrupole resonance, p. 215.

CHAPTER I

INTRODUCTION

1.1. Magnetic properties of nuclei

It is well known that many atomic nuclei possess an intrinsic angular momentum or spin. The largest measurable component of this angular momentum is $I\hbar$, where I, the nuclear spin number, is either an integer or a half-integer, and $\hbar$ is equal to $h/2\pi$, where h is Planck's constant. The electron, which has an intrinsic angular momentum of $\frac{1}{2}\hbar$, is known to have a magnetic moment of almost exactly one Bohr magneton, or $e\hbar/2M_e c$, where e is the electronic charge in electrostatic units, M_e is the electronic mass and c is the velocity of light. If we naïvely consider this magnetic moment to arise from the effective current produced by the rotating electronic charge, then a straightforward classical calculation shows that the value of the moment should indeed be of the order of a Bohr magneton. It is therefore to be expected that a magnetic moment should accompany the spin of an atomic nucleus, and that the value of this moment should be of the order of a quantity called the nuclear magneton, $e\hbar/2M_p c$, where M_p is the mass of the proton. This expectation is in fact realized, although the moments are never integral multiples of this unit. We may notice that since the mass of the proton is 1836 times larger than that of the electron, the nuclear magnetic moments are very small compared with the magnetic moment of the electron.

The properties of possessing an angular momentum and a magnetic moment were first ascribed to the nucleus by Pauli (1924) in order to account for the hyperfine structure of atomic spectra. He suggested that the structure originated in the interaction between the postulated nuclear magnetic moment and the orbital electrons. Advances, both in spectroscopic technique and in the theoretical elucidation of the spectra, have made it possible to determine the spin and magnetic moment of many nuclei. However, the accuracy of measurement of the magnetic moment by this method is not great.

In the three decades which have elapsed since Pauli put forward his suggestions, the concepts of nuclear spin and nuclear magnetic moment have found an important place in many other branches of physics. Of these we may mention in passing the following: molecular spectroscopy, in the Raman, infra-red and microwave regions; hyperfine structure of electronic paramagnetic resonance spectra; polarization of resonance radiation; nuclear scattering; excited states of nuclei; specific heats and *ortho-para* conversion of the two forms of hydrogen. In addition, the magnetic moment of the nucleus enters more directly into certain other branches of physics which we shall briefly discuss in the remaining sections of this introductory chapter.

1.2. Nuclear paramagnetism

Material which is otherwise diamagnetic exhibits a weak paramagnetism if the nuclei within it have a magnetic dipole moment. This nuclear paramagnetism was first directly demonstrated by Lasarew and Schubnikow (1937) for solid hydrogen. They measured the static magnetization by the Gouy method between 1·76 and 4·22° K. The paramagnetic contribution was distinguished from the diamagnetic contribution by its temperature variation. The experiment yielded a value for the proton magnetic moment with an accuracy of about 10%.

This work was important because it demonstrated the existence of nuclear magnetism by a conventional magnetic method using matter in bulk. Furthermore, the experiment showed that the nuclear magnets attained thermal equilibrium in the applied magnetic field at temperatures of the order of 2° K. in a short interval of time.

1.3. Molecular beams

A magnetic dipole experiences a force when placed in an inhomogeneous magnetic field. Atoms or molecules which possess a magnetic moment are therefore deflected on passing through such a field. This technique was used with great success in the classic atomic beam experiments of Stern and Gerlach (Stern, 1921; Gerlach and Stern, 1924) to prove experimentally that the measurable values of the component of an atomic magnetic moment do not form a continuous range; instead, they form a discrete set corresponding to the space quantization of the atom in the magnetic field. From the magnitude of the deflexion of the beam these workers were able to evaluate the atomic magnetic moment.

The beam-deflexion method was extended by Estermann and Stern (1933) and Frisch and Stern (1933) to measure the much smaller nuclear magnetic moment of the proton, using a beam of hydrogen molecules. The accuracy of the value obtained was about 10%.

The beam method was powerfully improved by Rabi and his colleagues (Rabi, Millman, Kusch and Zacharias, 1939) by the introduction of the resonance method. In addition to the steady magnetic field, the molecules of the beam are subjected to electromagnetic radiation of just such a frequency as to induce transitions between their quantized energy levels by a process of absorption or stimulated emission of quanta of energy. If the nuclear spin number is I, each energy level is split by the steady magnetic field into $(2I+1)$ equally spaced sub-levels. If the maximum measurable component of the nuclear magnetic moment is μ, the separation between the lowest and the highest sub-levels is $2\mu H$ in a steady magnetic field H; roughly speaking, these two levels correspond respectively to alignment of the nuclear moment with and against the magnetic field. The separation between successive sub-levels is therefore $\mu H/I$. The frequency of the electromagnetic radiation whose quanta can excite transitions between these sub-levels is thus equal to $\mu H/Ih$. The frequency of this resonant exchange of energy in a given field is found experimentally by the sharp reduction in the number of molecules reaching the detector; it is arranged that only molecules that suffer no change of energy have the correct deflexion in the inhomogeneous magnetic field to reach the detector.

In magnetic fields of several kilogauss the frequency of this nuclear magnetic resonance, as it is called, falls in the convenient radiofrequency range 10^5 to 10^8 c./s. The resonances are usually sharp, and enable the magnetic moment of nuclei to be obtained with an accuracy of a few parts in 10^4.

Although these elegant molecular beam experiments furnished the first successful example of nuclear magnetic resonance, we shall not be further concerned with this interesting type of work, which is a subject in itself. In this book our attention will be entirely devoted to nuclear magnetic resonance in matter in bulk.

1.4. Nuclear magnetic resonance in bulk matter

The resonant exchange of energy between the $(2I+1)$ energy levels of a nuclear magnetic moment in a magnetic field should not be restricted to matter in the form of molecular beams, but should also occur for matter in its ordinary solid, liquid or gaseous states. This had in fact been pointed out earlier by Gorter (1936), who unsuccessfully sought the resonance for the ^{7}Li nuclei in crystalline lithium fluoride, and for the protons in crystalline potassium alum. Both these nuclei have spin number $\frac{1}{2}$. Both therefore have two energy levels in a magnetic field, corresponding to alignment of the magnetic moment with or against the field. If the nuclear moments are in thermal equilibrium with the crystal lattice in which they are embedded, the number of nuclei in the lower energy level should be rather greater than the number in the upper level. Thus although the probability of transitions upwards (by absorption of a radio-frequency photon) equals the probability of transitions downwards (by stimulated emission of a photon), there should be a net absorption of energy by the nuclei. In the experiment just mentioned Gorter (1936) attempted to detect this absorption of energy by a calorimetric method.

Six years later Gorter and Broer (1942) attempted to find the nuclear magnetic resonance absorption for ^{7}Li in solid lithium chloride and for ^{19}F in solid potassium fluoride by looking for the anomalous dispersion which should accompany the absorption. Again a negative result was obtained. As Gorter (1951) has pointed out, the failure of both these attempts was mainly due to the use of unfavourable materials. For reasons which will become evident in Chapter 2, the applied radiofrequency field can, under certain conditions, bring about an equalization of the populations of the two energy levels, thus removing the difference in population upon which the observation of the effect depends.

The first successful nuclear magnetic resonance experiments using bulk material were carried out independently at the end of

1945 by Purcell, Torrey and Pound (1946) and by Bloch, Hansen and Packard (1946 a). Purcell and his colleagues found first of all the resonance absorption for the protons in solid paraffin, and detected it by measuring the additional loss which it caused in the tuned electrical circuit supplying the radiofrequency power. For this purpose the tuned circuit was placed in one arm of a balanced radiofrequency bridge. Bloch and his colleagues found their first resonance for the protons in water. Their method of detection was a novel one. The reorientation of the nuclear moments by the resonant electromagnetic field induces an electromotive force at the resonant frequency in a coil, whose axis is perpendicular to the steady magnetic field, wound round the phial of water. The manifestation of this electromotive force indicated the condition of resonance. Bloch has coined the name 'nuclear induction' for this method of detection.

As the fuller discussion in Chapter 2 will show, these two methods of detection are not based on fundamentally different principles. The names 'nuclear induction' and 'nuclear magnetic resonance' could therefore be applied to either.

1.5. Applications of nuclear magnetic resonance

The pioneer experiments discussed in the preceding section have opened up a rich field of research, and already in 1954, more than 400 publications record the work that has been done. The rapidity of the advance is partly to be explained by the relatively simple apparatus which the experiments require; even quite modestly equipped laboratories can enter the field.

The most obvious application is to the determination of nuclear properties. From the resonance condition one may obtain with considerable accuracy values of the nuclear gyromagnetic ratio† $\mu/I\hbar$ for any stable nucleus of reasonable abundance (§4.1). The nuclear spin number can often be found if it is not already known (§4.3), and hence the nuclear magnetic moment is obtained. The sign of the nuclear moment is also obtainable (§4.2). Nuclei with spin number $I \geqslant 1$ have an electric quadrupole moment, relative values

† There is much to be said for the nomenclature 'magnetogyric ratio' used by Bloembergen (1948). However, we use the more conventional, if less appropriate, name here.

of which may be determined from the fine structure of the nuclear magnetic resonance spectrum (§4.4 and Chapter 8).

Once a nuclear gyromagnetic ratio is known, it may be used for the calibration of magnetic fields, and for related problems with magnets (§§4.9, 4.10). The great precision of measurement of the nuclear magnetic resonance frequency in a given magnetic field has led to the accurate determination of certain important physical quantities (§§4.5–4.8), which in turn have necessitated a re-evaluation of the fundamental physical constants.

The nuclear magnetic resonance condition has been found in matter in all its forms: liquids and gases (Chapter 5), ionic and molecular solids (Chapter 6), and metals (Chapter 7). In each case, and particularly for the solid state, valuable information has been obtained concerning the structure and other non-nuclear properties of the material.

The resonance spectrum for liquids provides a tool for chemical analysis and identification (§§5.6, 5.7). Changes in this spectrum and of other resonance characteristics can be used to follow the progress of certain chemical reactions (§5.9). Information can be obtained concerning self-diffusion coefficients in liquids (§§5.2, 5.5) and the formation of complexes (§5.4). In suitable cases the water content of biological materials may be measured (§5.8).

The resonance spectrum of non-metallic solids frequently provides structural information (§§6.2, 6.3), and can lead to knowledge of hindered rotation of molecules or groups in the crystal (§6.4). Self-diffusion may be studied both in non-metals (§6.5) and in metals (§7.4). The protons in hydrated paramagnetic and antiferromagnetic crystals provide magnetic indicators for study of the magnetic properties of these crystals (§6.7.4). In the case of metals an interaction between the nuclei and the conduction electrons leads to information concerning the electronic states at the top of the Fermi distribution (§7.3). The resonance spectrum sometimes provides information concerning short-range order in alloys (§7.4), and concerning defects in both metals (§7.4) and non-metallic solids (§§4.3.1, 8.3).

Before discussing these applications, however, we shall first need to consider the theory of the phenomenon in more detail (Chapter 2), and also the experimental methods for its observation (Chapter 3).

BASIC THEORY

2.1. The resonance condition†

First of all let us consider an isolated nucleus in a steady magnetic field H_0. We will suppose that the nuclear spin number I is greater than zero, so that the nucleus may possess a magnetic moment. From the theory of quantum mechanics we know that the length of the nuclear angular momentum vector is $[I(I+1)]^{\frac{1}{2}}\hbar$, but that the only measurable components of this vector are given by $m\hbar$, where m, the magnetic quantum number, may take any of the $(2I+1)$ values in the series $I, I-1, I-2, ..., -(I-1), -I$. This is illustrated for the case of $I=\frac{3}{2}$ in fig. 1, where the angular momentum has just four measurable values along the direction of the applied magnetic field H_0.

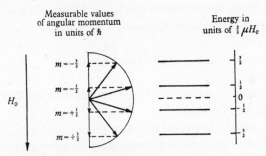

Fig. 1. Diagram showing schematically the four measurable values of angular momentum for a nucleus having spin number $I=\frac{3}{2}$, when placed in a magnetic field H_0. The corresponding four energy levels are also shown.

Corresponding to this quantization of the angular momentum components, the nuclear magnetic moment also has $(2I+1)$ components in proportion. In Chapter 1 we defined μ as the maximum measurable component of the magnetic moment. It must therefore be remembered that the length of the magnetic moment vector is actually $[(I+1)/I]^{\frac{1}{2}}\mu$. However, μ is the quantity which is of phys-

† The approach in the earlier part of this chapter is based upon sections of the classic paper by Bloembergen, Purcell and Pound (1948).

ical interest, and for this reason experimental workers usually derive and state the value of μ from their measurements, and frequently refer to this quantity simply as the 'magnetic moment'. The components of the magnetic moment are given by the $(2I+1)$ values of $m\mu/I$, forming the series $\mu, (I-1)\mu/I, ..., -(I-1)\mu/I, -\mu$. The energy levels of the nuclear magnet in the magnetic field $\mathbf{H_0}$ are therefore given by the $(2I+1)$ values of $-m\mu H_0/I$. These levels also are illustrated in fig. 1 for the case of $I=\frac{3}{2}$, and in the general case show a set of equally spaced levels with separation $\mu H_0/I$ between successive levels. This energy separation is often written as $g\mu_0 H_0$, where μ_0 is the nuclear magneton, and $g\,(=\mu/\mu_0 I)$ is called the *splitting factor* or *g-factor*. This g-factor is in fact the counterpart of the Landé splitting factor in atomic spectroscopy. It will be noticed that the quantity gI is the magnetic moment measured in units of the nuclear magneton.

The selection rule governing transitions between energy levels is the same as for the closely related Zeeman effect; transitions are allowed which cause m to change by ± 1. A quantum of energy can therefore excite transitions between the energy levels if it has the same magnitude as the level spacing:

$$h\nu_0 = \frac{\mu}{I} H_0 = g\mu_0 H_0, \tag{2.1}$$

where ν_0 is the frequency of the electromagnetic radiation supplying the quanta of energy.

For the proton the value of g is approximately 5·58. Thus, in a typical field of 5000 gauss the resonance frequency given by (2.1) is 21·3 Mc./s. Only the triton ^{3}H has a greater g-factor, corresponding to a frequency of 22·7 Mc./s. in this field, while for most other nuclei which possess a magnetic moment the resonance frequency is greater than 1 Mc./s. in the same field. The frequencies thus fall in a convenient radiofrequency band.

In conventional optical spectroscopy emitted radiation is analyzed, whereas in the nuclear magnetic resonance absorption experiment, as also for molecular beam magnetic resonance, radiation is generated externally and its effect on the atomic system is investigated. Notwithstanding this difference in technique, we may find the necessary condition for observation of the magnetic resonance transitions from the properties of the emission spectrum of the analogous atomic Zeeman effect. Zeeman effect transitions which involve a change in m of ± 1 produce radiation which is circularly

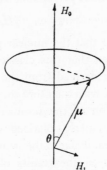

Fig. 2. Diagram illustrating the classical Larmor precession of a magnetic moment μ in a magnetic field H_0.

polarized in the plane perpendicular to the steady magnetic field H_0. In order to excite such transitions in the magnetic resonance experiment, it is therefore necessary to supply radiation with the magnetic vector circularly polarized in a plane perpendicular to the steady magnetic field.

This requirement of circular polarization is just what one would expect by classical argument. If a magnetic dipole μ is placed in a magnetic field H_0 as shown in fig. 2, the dipole precesses about the direction of the applied field. The rate of precession is given by the well-known Larmor angular frequency

$$\omega_0 = \gamma H_0, \tag{2.2}$$

where γ is the gyromagnetic ratio of the dipole. Suppose now that an additional small magnetic field H_1 is applied at right angles to H_0, in the plane containing μ and H_0. The dipole will experience a couple $(\mu \wedge H_1)$ tending to increase the angle θ between μ and H_0. If the small field H_1 is made to rotate about H_0 as axis in synchronism with the precession of the dipole, this couple will cause the angle θ to increase steadily. If, on the other hand, H_1 rotates with an angular frequency different from the Larmor precessional frequency, or in opposite sense, the couple $(\mu \wedge H_1)$ will vary in magnitude and direction according to the relative phases of the two motions, and will merely produce small perturbations of the precessional motion with no net effect. A resonance therefore occurs when the angular frequency $2\pi\nu$ of the rotating field is equal to the angular frequency of Larmor precession, namely, when

$$2\pi\nu = 2\pi\nu_0 = \omega_0 = \gamma H_0. \tag{2.3}$$

When it is remembered that the nuclear gyromagnetic ratio γ is given by $\mu/I\hbar$, it is seen that this classical resonance condition agrees exactly with that derived from the quantum theory (2.1). Moreover, we see that as with quantum theory, so also classically, a condition for observation of the resonance is that the electromagnetic radiation be circularly polarized with the magnetic vector rotating in a plane perpendicular to the steady magnetic field.

The agreement between the results of the classical and the quantum theory viewpoints allows a number of features of the nuclear magnetic resonance phenomenon to be discussed rather simply in terms of a classical vector model of the nucleus.

Although the generation of a high-frequency rotating magnetic field is quite practicable (see §4.2), it is usually much simpler to provide a linearly oscillating field. Fortunately, for most purposes

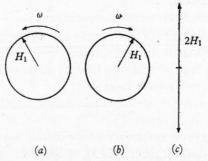

(a)   (b)   (c)

Fig. 3. If the two equal vectors H_1, shown in (a) and (b) rotating in opposite senses, are superimposed, the resultant (c) is a linear vibration of amplitude $2H_1$. The steady field $\mathbf{H}_0$ is taken normal to the diagram.

linear polarization is quite adequate, since, as in the theory of rotatory polarization in optically active crystals, a linearly oscillating field may be regarded as the superimposition of two rotating fields. Thus, as shown in fig. 3, if the linearly oscillating field has amplitude $2H_1$, it may be decomposed into two circularly polarized fields, each of amplitude H_1, but rotating in opposite senses in a plane perpendicular to H_0. Resonance will be obtained with the component which has the correct sense, the other component having negligible effect. It is not generally possible to tell which component is utilized, and it is not usually necessary to know this. The information is only of importance when it is desired to find the sign of μ; from (2.2) one sees that it is the sign of the gyromagnetic ratio γ, and therefore of μ, which determines the sense of the Larmor precession. This question is discussed more fully in §4.2.

2.2. Spin-lattice relaxation time

Let us now consider an assembly of identical atomic nuclei in the presence of a steady magnetic field H_0. For simplicity we will at first suppose that the nuclear spin number is $\frac{1}{2}$. It will further be assumed that there is only a weak coupling between the nuclear magnets. This allows us to neglect to a first approximation the magnetic interaction between the nuclei, and so to take the energy levels discussed in the previous section for an isolated nucleus as those for each nucleus in the assembly. At the same time some coupling between the nuclei has to be assumed so that the assembly may be considered to be in thermal equilibrium at a temperature T_s.

Since we are concerned with resonance in matter in its normal physical and chemical states, the assembly of nuclei must be present in their usual role as central particles in atomic systems. We shall, however, suppose that the interaction of the nuclear magnets with the remainder of the system is even smaller. If this assumption is to be reasonable we must exclude for the present atomic and molecular systems which possess a permanent magnetic dipole moment; in fact, apart from the feeble paramagnetism produced by the nuclei, the material concerned must be diamagnetic. The material in which the nuclear magnets are embedded is generally referred to as the 'lattice', whether it be solid, liquid or gas.

Since $I = \frac{1}{2}$, each nucleus has two possible energy levels separated by a gap of $2\mu H_0$. If we now apply radiation of the resonant frequency polarized in a direction perpendicular to H_0, transitions between the two levels take place. From the simple theory of the Einstein coefficients (Einstein, 1917) we know that the probability of transitions upwards by absorption is equal to the probability of transitions downwards by stimulated emission. In comparison with these probabilities, the probability of transitions downwards by spontaneous emission is quite negligible (Purcell, 1946). If the numbers of nuclei in each energy level were equal, the average rate of transitions up and down would therefore be equal, and there would be no net effect on the system. Actually, however, since the nuclear spins are in equilibrium at temperature T_s, the population of the lower level exceeds that of the upper level by the Boltzmann factor $\exp(2\mu H_0/kT_s)$, where k is Boltzmann's constant. At room temperature, for protons in a field of 5000 gauss, this factor has the value

$$\exp\left[\frac{2\mu H_0}{kT_s}\right] \simeq 1 + \frac{2\mu H_0}{kT_s} = 1 + 4 \times 10^{-6}. \tag{2.4}$$

On account of this typically small, but finite, excess of population in the lower energy state, there is a net absorption of energy from the radiofrequency field.

The absorption of energy corresponds to the transfer of some of the excess population in the lower level to the upper level. If there is no interaction between the system of nuclear spins and the lattice, the fractional excess of population, $2\mu H_0/kT_s$, steadily dwindles. The ratio of the populations may still formally be described by a Boltzmann factor, with the temperature T_s of the spin system steadily rising. The spin system is in fact being subjected to a process of radiofrequency heating. The temperature of the lattice is not affected however, since we have assumed negligible interaction between the spin system and the lattice.

In any actual physical system the nuclear spins are never entirely bereft of interaction with the lattice, although very often the interaction is quite small. Indeed, it is only because we have assumed, as is frequently the case, that the interactions within the spin system and within the lattice are considerably greater than the interaction between the two systems that we are justified in speaking separately of a 'spin temperature' and a 'lattice temperature'. Such interaction as there is between the two systems tends to bring both into thermal equilibrium at the same temperature. In practice this common temperature is almost identical with the lattice temperature, since except at extremely low temperatures the heat capacity of the spin system is very small compared with that of the lattice. Thus, while the radiofrequency radiation is reducing the excess of population in the lower energy state, the interactions with the lattice are tending to restore the excess to its original value.

Suppose the spin temperature is initially equal to that of the lattice, and we raise it to a value greater than the lattice temperature by application of the radiofrequency field. If this field is now removed, we may ask how rapidly will the temperature of the spin system return to that of the lattice?

The same question arises also in another way. Let us suppose our spin system and lattice are in thermal equilibrium at a common temperature in absence of the radiofrequency field and in a vanishingly small steady magnetic field. The two energy levels are scarcely separated and their populations are practically identical. Suppose now we suddenly apply a steady magnetic field of several kilogauss. In this field the almost identical populations correspond to a very high spin temperature, and we ask how rapidly will the populations adjust themselves, by interaction with the lattice, to the values required by thermal equilibrium of spins with the lattice?

In order that the spin system may cool down to the temperature of the lattice there must be a net downward transfer of nuclei to the lower energy level. Since the population of the upper level does not exceed that of the lower level, this can only be possible if the probability for downward transitions exceeds that for upward transitions. This may at first sight appear to contradict the earlier statement that the two transition probabilities are equal. It must, however, be remembered that the earlier statement referred to transitions stimulated by interaction with the radiofrequency field, whilst here we are dealing with the more complicated interaction of the spin system with the lattice. Indeed, we can show that in this case the probability W_- per unit time for downward transitions must exceed the probability W_+ for upward transitions.

Let the number of nuclei per cm.[3] in the upper energy state ($m = -\frac{1}{2}$) be N_-, and in the lower state ($m = +\frac{1}{2}$) be N_+. If the whole system of spins and lattice were in thermal equilibrium at a temperature T, then by the principle of detailed balancing, the number of transitions upwards and downwards must be equal, which gives

$$W_+ N_+ = W_- N_-.$$

In equilibrium N_+ and N_- are related by the Boltzmann factor, so that

$$\frac{W_-}{W_+} = \frac{N_+}{N_-} = \exp\left[\frac{2\mu H_0}{kT}\right] \cong 1 + \frac{2\mu H_0}{kT}. \tag{2.5}$$

We may therefore write

$$\left. \begin{array}{l} W_- = W\left(1 + \dfrac{\mu H_0}{kT}\right) \\[2mm] W_+ = W\left(1 - \dfrac{\mu H_0}{kT}\right) \end{array} \right\}, \tag{2.6}$$

and

where W is the mean of the two transition probabilities W_+ and W_-.

Bloembergen (private communication) has indicated the physical reason for this difference in upward and downward probability rates by means of a simple model. Suppose the lattice is simulated by a set of harmonic oscillators whose equally spaced energy levels are separated by $2\mu H_0$, the nuclear energy level difference. The probability that a nucleus makes an upward transition by exchanging a quantum of energy $2\mu H_0$ with a given oscillator is proportional to the probability that that oscillator is in a state other than the lowest energy state, since in this lowest state it cannot give up a quantum of energy; this probability is $\left[1 - \dfrac{2\mu H_0}{kT}\right]$, since, as one can easily show, the probability of occupation of the lowest state is just $2\mu H_0/kT$. On the other hand the oscillator can *receive* a quantum of energy whatever level it is occupying. The ratio of the probability of surrendering a quantum to a nucleus to the probability of receiving one from the nucleus is therefore $\left[1 - \dfrac{2\mu H_0}{kT}\right]$, and this consequently is the ratio of the probabilities of a nucleus making upward and downward transitions by interaction with the oscillator, in agreement with (2.5). It can be shown that other models lead to the same relation.

If now we consider our spin system at a temperature T_s different from the lattice temperature T, and remember that the excess number n of nuclei per cm.[3] in the lower state, where $n = N_+ - N_-$, changes by 2 with each transition, then the rate of change of n is given by

$$\frac{dn}{dt} = 2N_-W_- - 2N_+W_+. \tag{2.7}$$

Using (2.6) this becomes

$$\frac{dn}{dt} = 2W(n_0 - n),\tag{2.8}$$

where
$$n_0 = N\mu H_0/kT,\tag{2.9}$$

and $N = N_+ + N_-$ is the total number of nuclei per cm.[3]. Thus n_0 is the value of n when the spin system is in thermal equilibrium with the lattice. Integration of (2.8) gives

$$n_0 - n = (n_0 - n_a)e^{-2Wt},\tag{2.10}$$

where n_a is the initial value of n. Equation (2.10) shows that equilibrium is approached exponentially with a characteristic time

$$T_1 = 1/2W.\tag{2.11}$$

The approach of the spin and lattice systems to equilibrium may be termed a *thermal relaxation process*, and for this reason, T_1 is called the *spin-lattice relaxation time* or the *thermal relaxation time*. Bloch (1946) has given T_1 the alternative name of *longitudinal* relaxation time. The reason for this nomenclature will become apparent in §2.7.

So far we have considered only the relaxation process for a system of nuclei having spin number $I = \frac{1}{2}$. If $I > \frac{1}{2}$, there are, of course, more than two energy levels. The spin system can only then be characterized by a spin temperature T_s, if the populations of the successive equally spaced levels are related by the same factor. Broer (1945) has shown that if this condition is met, and if T_s differs from the lattice temperature T, the approach to equilibrium is similar to that for $I = \frac{1}{2}$. A temperature can be assigned to the system at any stage in the relaxation process, and the excess number in each level with respect to the level next above changes with the same characteristic time T_1, following equations (2.10) and (2.11).

The nature of the spin-lattice interaction and the theoretical evaluation of the relaxation time T_1 is discussed in later sections (§§5.2, 6.6, 7.2, 8.4). It may be mentioned here, however, that the values of T_1 which are encountered experimentally usually lie within the range 10^{-4} to 10^4 sec. The value is usually longer for solids than for liquids and gases. For solids it is rarely shorter than 10^{-2} sec., while at low temperatures it may be very long. For pure liquids T_1 may be as short as 10^{-2} or 10^{-3} sec., and rarely exceeds 10 sec. The presence of paramagnetic ions in a liquid promotes the relaxation process, and may reduce T_1 to less than 10^{-4} sec. In general the value of T_1 also depends on the strength H_0 of the steady field.

2.3. Spin-spin interaction

We have so far been concerned only with the interaction between the nuclear-spin system and the lattice, and we must now examine the effect of interactions between the spins themselves. Since each nucleus possesses a small magnetic dipole moment there will be a magnetic dipole-dipole interaction between each pair of nuclei. From a classical point of view this may be regarded in the following way. Each nuclear magnet finds itself not only in the applied steady magnetic field $\mathbf{H_0}$, but also in a small local magnetic field $\mathbf{H}_{local}$ produced by the neighbouring nuclear magnets. The direction of the local field differs from nucleus to nucleus, depending on the relative disposition of the neighbouring nuclei in the lattice, and on their magnetic quantum number m.

The magnetic field of a magnetic dipole of moment μ at a distance r is of the order of μ/r^3. The field therefore falls off rapidly with increase of r, so that only the nearest neighbours make important contributions to $\mathbf{H}_{local}$. Taking for r a value of 1 Å, and for μ one nuclear magneton, we find

$$H_{local} \sim \mu_0 r^{-3} \sim 5 \text{ gauss.} \tag{2.12}$$

The steady magnetic field will not therefore be the same for each nucleus, but will vary over a range of several gauss from one nucleus to another. It follows that the resonance condition will not be perfectly sharp as was suggested for a system of very weakly interacting spins by equation (2.1). Instead, the energy levels are broadened by an amount of order $g\mu_0 H_{local}$. If we have a fixed radiofrequency ν_0 and traverse the steady magnetic field, the resonance will be found to be spread about H_0 over a range of values of the order of H_{local}.

Since the steady field differs from nucleus to nucleus, there will be a distribution of the frequencies of their Larmor precession, covering a range $\delta\omega_0$, which is found from (2.2) and (2.12) to be

$$\delta\omega_0 \sim \gamma H_{local} \sim \mu_0^2/\hbar r^3 \sim 10^4 \text{ sec}^{-1}. \tag{2.13}$$

If two spins have precession frequencies differing by $\delta\omega_0$ and are initially in phase, then they will be out of phase in a time $\sim 1/\delta\omega_0$, namely, about 10^{-4} sec.

This is not the only way in which the relative phases of two precessing spins j and k may be disturbed. Nucleus j produces at nucleus k a magnetic field oscillating at its Larmor frequency, and as a result may induce a transition in k. The energy for the transition of nucleus k comes of course from nucleus j; there is, in fact, a mutual exchange of energy in the process. Since the relative phases of the nuclei change in a time of the order of $1/\delta\omega_0$, the correct phasing for this spin-exchange process should occur after a time interval of this order, and this in turn should determine the lifetime of a spin state. It is well known that a lifetime of $1/\delta\omega_0$ produces a broadening of the energy levels of $\sim\hbar\delta\omega_0$. From (2.13) and the definition of γ, this energy breadth is

$$\hbar\delta\omega_0 \sim \hbar\gamma H_{\text{local}} = g\mu_0 H_{\text{local}}. \tag{2.14}$$

It therefore follows from the resonance condition (2.1) that this process results in a further broadening of the resonance line at fixed radiofrequency by an amount of the order H_{local}.

These two phase-disturbing and line-broadening processes are only both present when identical nuclei are concerned. For a system of non-identical nuclei, the local field effect is still present, but the spin-exchange process is absent, since the Larmor precession frequencies of the nuclei are quite different.

It is convenient to introduce a spin-spin interaction time T_2 to describe the lifetime or phase-memory time of a nuclear spin state, where $T_2 \sim 1/\delta\omega_0 \sim 10^{-4}$ sec. A more precise definition of T_2 is given later in this section. Bloch (1946) has called T_2 the *transversal relaxation time*. The reason for this name will become apparent in §2.7.

The quantum-mechanical treatment of the spin-spin interaction is deferred to Chapters 5 and 6, where it is found that the simple order of magnitude arguments given here are substantially correct. A preliminary mention should, however, be made here that for liquids and gases the reorientation and diffusion of the molecules is usually so rapid that the local magnetic field is smoothed out to a very small average value, yielding quite a narrow resonance line. This effect is found in some solids also.

Although we have considered here only the basic broadening of the resonance line due to nuclear spin-spin interaction, we must not overlook the possibility of other sources of broadening. In particular we may mention: (*a*) non-uniformity of the steady magnetic field over the assembly of nuclear spins, to which reference will again be made in §3.9.2; (*b*) broadening caused by a very short spin-lattice relaxation time T_1 limiting the lifetime of the nucleus in a given state, which is discussed in §5.3; (*c*) nuclear electric quadrupole interaction if the spin number I exceeds $\frac{1}{2}$, which is discussed in Chapter 8.

Let us now describe the shape of the absorption line as a function of frequency in a fixed magnetic field H_0 by means of a normalized function $g(\nu)$, where ν is frequency and

$$\int_0^\infty g(\nu)\, d\nu = 1. \tag{2.15}$$

For absorption lines of similar shape, the maximum value of the shape function, $g(\nu)_{\text{max}}$, will be large for narrow lines and small for broad lines. Thus the quantity $1/g(\nu)_{\text{max}}$ is a rough guide to the width of the line. From our previous discussion the line width reckoned as a function of frequency is of the order of $\delta\omega_0$; hence the spin-spin interaction time T_2 is of the order of $g(\nu)_{\text{max}}$. We use this connexion to form a precise definition of T_2, and following Bloembergen, Purcell and Pound (1948), we put

$$T_2 = \tfrac{1}{2}g(\nu)_{\text{max}}. \tag{2.16}$$

The coefficient $\frac{1}{2}$ is inserted to make this definition of T_2 consistent with that of Bloch (1946) (see §2.7). From the definition (2.16), T_2 is connected only with the peak value of the normalized line-shape curve.

2.4. Saturation

In §2.2 we considered the rate at which the spin system and the lattice approach thermal equilibrium with each other in a steady magnetic field. We will now consider the condition of the system after settling down to a steady state under the influence of applied electromagnetic radiation. We will assume for simplicity that $I = \frac{1}{2}$.

In absence of radiation, the differential equation governing the time variation of the excess number n of nuclei per cm.³ in the lower state, is found from (2.8) and (2.11) to be

$$\frac{dn}{dt} = \frac{n_0 - n}{T_1}.\qquad(2.17)$$

When radiation is present another term must be added to account for the upward transitions which correspond to the net absorption of energy. Thus we get

$$\frac{dn}{dt} = \frac{n_0 - n}{T_1} - 2nP,\qquad(2.18)$$

where P is the probability per unit time of a transition by a nucleus between the two levels under the influence of the radiation (each upward transition reduces n by 2). A steady state is reached when $dn/dt = 0$, and in this condition, using (2.18), the steady state value n_s of the excess number is given by

$$\frac{n_s}{n_0} = \frac{1}{1 + 2PT_1}.\qquad(2.19)$$

It is now necessary to evaluate the transition probability P. If a nucleus is subjected to a radiofrequency magnetic field of amplitude H_1 rotating in the correct sense in a plane perpendicular to $\mathbf{H}_0$, the probability of a transition in unit time between states designated by magnetic quantum numbers m and m' is found by standard radiation theory to be (Bloembergen, Purcell and Pound, 1948):

$$P_{m \to m'} = \tfrac{1}{2}\gamma^2 H_1^2 \, | \langle m \, | \, I \, | \, m' \rangle \, |^2 \, g(\nu),\qquad(2.20)$$

where $\langle m \, | \, I \, | \, m' \rangle$ is the appropriate matrix element of the nuclear spin operator.

The non-diagonal matrix elements $\langle m \, | \, I \, | \, m' \rangle$ are zero except when $|m - m'| = 1$, corresponding to the selection rule mentioned in §2.1. For $m' = m - 1$ we have (see Condon and Shortley, 1935, §3³)

$$| \langle m \, | \, I \, | \, m' \rangle \, |^2 = \tfrac{1}{2}(I + m)(I - m + 1).\qquad(2.21)$$

Combining equations (2.20) and (2.21) we get

$$P_{m \to m-1} = \tfrac{1}{4}\gamma^2 H_1^2 g(\nu)(I + m)(I - m + 1).\qquad(2.22)$$

For the case $I = \tfrac{1}{2}$, this reduces to

$$P = \tfrac{1}{4}\gamma^2 H_1^2 g(\nu).\qquad(2.23)$$

This result may now be combined with (2.19) to give

$$n_s/n_0 = [\mathrm{I} + \tfrac{1}{2}\gamma^2 H_1^2 T_1 g(\nu)]^{-1} \equiv Z. \tag{2.24}$$

If a radiofrequency field is applied, whose amplitude H_1 is large, n_s/n_0 becomes quite small; the spin temperature T_s becomes very high, and the spin system is said to be *saturated*. For this reason the expression on the right-hand side of (2.24), to which we have given the symbol Z, is called the *saturation factor*. Saturation is greatest at the frequency which gives $g(\nu)$ its maximum value. Using (2.16) the saturation factor then has the value

$$n_s/n_0 = [\mathrm{I} + \gamma^2 H_1^2 T_1 T_2]^{-1} = Z_0, \tag{2.25}$$

where Z_0 is the value of the saturation factor at the maximum of the line-shape function $g(\nu)$.

The excess number n_0, when the spin system and the lattice are in thermal equilibrium, is related to the common spin and lattice temperature T by (2.9). The general excess number n is related to the spin temperature T_s by the similar equation

$$n = N\mu H_0/k T_s. \tag{2.26}$$

With (2.9) this gives the spin temperature of the steady state as

$$T_s = T n_0/n_s = T/Z. \tag{2.27}$$

The spins can quite readily be heated up to extremely high temperatures. For example, the spin-lattice relaxation time T_1 for the protons in ice in a field of 7000 gauss at $88°$ K. is of the order of 10^4 sec. (Sachs and Turner, 1951), while T_2 is of the order of 10^{-5} sec. If the system is subjected to a radiofrequency magnetic field having amplitude $H_1 = 0 \cdot 1$ gauss, a value which is easily realized, then the saturation factor Z_0 is of the order 10^{-6}, and the spin temperature when a steady state has been reached is about 10^8 °K.! These very high temperatures, which are attained in a second or so, do not of course require the absorption of much energy from the radiofrequency field. In order to raise the spin temperature from that of the lattice, where the excess population of the lower state is n_0, to an infinite temperature, where the excess is exactly zero, energy must be supplied per cm.^{3} equal to $n_0\mu H_0$. Using (2.9) this energy is given by $N(\mu H_0)^2/kT$. Applying this result to the above-mentioned case of ice at $88°$ K., the necessary energy is found to be about 1 erg per mole or about 2×10^{-8} calorie per mole. In passing we may mention, as is discussed in §6.7.1, that still more remarkable, apparently negative, values of the spin temperature can be attained.

We will now consider the rate of approach of the spin system to the steady state given by (2.24). This approach is governed by the differential equation (2.18), whose solution is

$$(n_s - n) = (n_s - n_a) \exp(-t/T_1 Z), \qquad (2.28)$$

where n_a is the initial value of n. The approach to the steady state in presence of radiation is thus of the same exponential form as that given by (2.10) when radiation is absent, but the characteristic time is now $T_1 Z$ instead of just T_1.

2.5. Magnetic susceptibilities

We have seen that an assembly of nuclear magnets in a steady magnetic field absorbs power from a suitably applied radio-frequency field. From a macroscopic point of view this absorption may be described by means of the imaginary part χ'' of the complex nuclear magnetic susceptibility, $\chi = \chi' - i\chi''$, of the assembly. However, before deriving this quantity in terms of the properties of the spin system, we will first derive the static magnetic susceptibility of the system, χ_0, which we shall need in later calculations.

Consider an assembly of identical weakly interacting nuclei of spin number I, in thermal equilibrium at a spin temperature T_s in a steady magnetic field $\mathbf{H}_0$. As we saw in §2.1, nuclei having magnetic quantum number m are to be found in the energy level $-m\mu H_0/I$. The population of the level is therefore weighted by the Boltzmann factor

$$\exp\left(\frac{m\mu H_0}{IkT_s}\right) \cong 1 + \frac{m\mu H_0}{IkT_s}. \qquad (2.29)$$

The approximation made in (2.29) is a very good one for all practical conditions. Hence the population of each level per cm.³, $N(m)$, is given by

$$N(m) = \frac{N}{2I+1}\left(1 + \frac{m\mu H_0}{IkT_s}\right), \qquad (2.30)$$

since the proportionality constant $N/(2I+1)$ is such that it gives correctly the total population in all $(2I+1)$ levels:

$$\sum_{m=-I}^{I} N(m) = N. \qquad (2.31)$$

The total magnetic moment per cm.^{3}, namely, the magnetization $\mathcal{M}$, is therefore given by

$$\mathcal{M} = \sum_{-I}^{I} N(m) \, m\mu/I = \frac{N\mu^2 H_0}{I^2(2I+1)k \, T_s} \sum_{-I}^{I} m^2$$

$$= \frac{N\mu^2 H_0(I+1)}{3kT_sI}, \qquad (2.32)$$

since $\sum_{-I}^{I} m^2 = \tfrac{1}{3}I(I+1)(2I+1).$

The static susceptibility is therefore given by

$$\chi_0(T_s) = \frac{\mathcal{M}}{H_0} = \frac{N\mu^2(I+1)}{3kT_sI}. \qquad (2.33)$$

We may note in passing that if we write $\mu' = [(I+1)/I]^{\frac{1}{2}}\mu$, where μ' is the length of the nuclear magnetic moment vector (see §2.1), then (2.33) agrees with the classical formula of Langevin (1905).

The static nuclear magnetic susceptibility given by (2.33) was the quantity measured by Lasarew and Schubnikow (1937) for solid hydrogen, as was mentioned in §1.2. Since this susceptibility is always very small, their work was done at low temperatures in order to make the most of the inverse dependence on temperature.

We now turn to the radiofrequency susceptibility. First of all we will find the power absorbed by the system in terms of χ''. Instead of generating a rotating magnetic field, we apply as usual a linearly oscillating magnetic field given by

$$H = 2H_1 \cos 2\pi\nu t. \qquad (2.34)$$

This field produces an in-phase magnetization $2\chi'H_1 \cos 2\pi\nu t$ and an out-of-phase component $2\chi''H_1 \sin 2\pi\nu t$. Now the mean rate of absorption of energy per unit volume is the mean value of $Hd\mathcal{M}/dt$. Only the out-of-phase component contributes to this mean, so that the power absorbed per cm.^{3}, A, is given by

$$A = \nu \int_0^{1/\nu} H \frac{d\mathcal{M}}{dt} \, dt = 8\pi\nu^2\chi''H_1^2 \int_0^{1/\nu} \cos^2 2\pi\nu t \, dt = 4\pi\nu\chi''H_1^2. \quad (2.35)$$

We must now find A in terms of the nuclear system. First consider the adjacent energy levels m and $m-1$. The excess number of nuclei in the lower of these two levels is found from (2.30) to be

$$N(m) - N(m-1) = \frac{N\mu H_0}{(2I+1) \, IkT_s} = \frac{Nh\nu_0}{(2I+1)kT_s}, \qquad (2.36)$$

using (2.1). Since each upward transition between these two levels requires the absorption of a quantum $h\nu$ of energy, and since the probability of such a transition is $P_{m \to m-1}$, the power absorbed by transitions between these two levels is

$$\left(\frac{Nh^2\nu_0\nu}{(2I+1)kT_s}\right)P_{m \to m-1}. \qquad (2.37)$$

The total power absorbed is obtained by summing (2.37) over all the energy levels. Using (2.22) for $P_{m \to m-1}$ we get

$$A = \tfrac{1}{4}\gamma^2 H_1^2 g(\nu)\left(\frac{Nh^2\nu_0\nu}{(2I+1)kT_s}\right)\sum_{I}^{-I+1}(I+m)(I-m+1). \qquad (2.38)$$

The sum is readily found to be $\tfrac{2}{3}I(I+1)(2I+1)$. Combining (2.38) with (2.35), and substituting $\gamma = \mu/I\hbar$, the imaginary part of the susceptibility is found to be

$$\chi'' = \frac{\pi}{2}\left[\frac{N\mu^2(I+1)}{3kT_sI}\right]\nu_0 g(\nu). \qquad (2.39)$$

The term in the square bracket is the static susceptibility $\chi_0(T_s)$ given by (2.33). When the spin temperature T_s is equal to the lattice temperature T, we shall write the static susceptibility as χ_0 without specifying the temperature. Since the spin and lattice temperatures are related by the saturation factor Z (2.27), it follows that

$$\chi_0(T_s) = Z\chi_0.$$

Equation (2.39) for χ'' may therefore be written as

$$\chi'' = \tfrac{1}{2}\pi\chi_0(T_s)\nu_0 g(\nu) = \tfrac{1}{2}\pi Z\chi_0\nu_0 g(\nu). \qquad (2.40)$$

The correctness of this result may be checked with the aid of the Kronig–Kramers relations. These relations were originally derived by Kronig (1926) and Kramers (1927) for the electric susceptibility. They are, however, applicable in many other branches of physics, and in particular were shown by Gorter and Kronig (1936) to be equally valid for the magnetic susceptibility. They express a connexion between the real and imaginary parts of the susceptibility. Applied to our problem these equations state:

$$\left.\begin{array}{l} \chi'(\nu) = \dfrac{2}{\pi}\displaystyle\int_0^\infty \dfrac{\nu'\chi''(\nu')d\nu'}{\nu'^2 - \nu^2}, \\[4mm] \chi''(\nu) = -\dfrac{2\nu}{\pi}\displaystyle\int_0^\infty \dfrac{\chi'(\nu')d\nu'}{\nu'^2 - \nu^2}. \end{array}\right\} \qquad (2.41)$$

C

For the real part of the susceptibility at zero frequency, the first of these equations, substituting for χ'' from (2.40), gives:

$$\chi'(0) = \chi_0(T_s)\nu_0 \int_0^\infty \frac{g(\nu')}{\nu'}\,d\nu'. \tag{2.42}$$

In a field $\mathbf{H}_0$, the shape function $g(\nu')$ is effectively zero except when the frequency ν' is close to the resonant frequency ν_0. Putting $\nu' \cong \nu_0$ in the denominator of the integral, and remembering from (2.15) that $g(\nu')$ is a normalized function, we see that $\chi'(0)$ does correctly reduce to $\chi_0(T_s)$, the static susceptibility.†

We may derive another piece of information from the first of equations (2.41). Since the integrand has a singularity at $\nu' = \nu$, the main contribution to the integral comes from the values of ν' near ν, when $\chi''(\nu') = \chi''(\nu)$. Moreover, $\chi''(\nu)$, which is proportional to the line-shape function $g(\nu)$, is small except when ν is close to the resonant frequency ν_0. It therefore follows from (2.41) that the real part also of the susceptibility only has appreciable values at frequencies close to ν_0. This variation of $\chi'(\nu)$ near the resonant frequency is just another example of dispersion accompanying absorption, which in this case is represented by $\chi''(\nu)$.‡

2.6. Conditions for observation of nuclear magnetic resonance absorption

We are now in a position to enumerate the basic requirements for observing nuclear magnetic resonance absorption. A sample of material, probably about 1 cm.³ in size, containing large numbers of the nuclei in which we are interested, is subjected to a steady magnetic field $\mathbf{H}_0$. It is then necessary to wait for a period several times longer than the spin-lattice relaxation time T_1 for the spin system to come into thermal equilibrium with the lattice in the steady field. Facilities must now be provided for subjecting the sample to a linearly polarized radiofrequency magnetic field in a direction at right angles to $\mathbf{H}_0$. This is readily achieved by winding

---

† In obtaining (2.42) it has been assumed that the spin temperature T_s remains the same at all frequencies. This implies either that there is no saturation ($Z = 1$) or that the degree of saturation remains constant throughout; if neither situation is true the Kronig–Kramers relations cannot be used, since they apply to systems which are not being saturated (Portis, 1953).

‡ For further discussion of this point see Pake and Purcell (1948).

a coil round the sample, its axis in the direction of the desired oscillatory field, and by passing a radiofrequency current through the coil. Provision must be made for adjustment either of the magnitude of the steady field $\mathbf{H}_0$, or of the radiofrequency ν, until the resonance condition is reached. In this condition the greatest power is absorbed from the radiofrequency field, and the imaginary part of the nuclear magnetic susceptibility χ'' has its peak value. The condition may be detected by observing the additional power loss, equivalent to extra resistance, in the coil supplying the radiofrequency. Alternatively, the change in reactance of the coil due to the dispersive change of χ' may be sought. A description of the experimental methods for making these measurements is deferred to Chapter 3.

We must note, however, that χ'' is proportional to the saturation factor Z given by (2.24). If measurements of the absorption are being made, it is therefore important that the amplitude $2H_1$ of the radiofrequency field shall not be too large, as otherwise the sample will become saturated and χ'' may become too small to measure. The absorption may of course be observable initially while the spin system is still in process of becoming saturated, but has not yet settled down to its steady state. However, since the time for the attainment of this steady state is $T_1 Z$ (2.28), a high degree of saturation is more rapidly reached than slight saturation.

One can thus fail to find nuclear magnetic resonance absorption either by looking for it too soon, or by looking too late. The first case applies if T_1 is long and insufficient time has elapsed to allow the spins and the lattice to come into equilibrium in the steady field. The second case applies if a large radiofrequency field is applied, giving a small saturation factor Z, and if observation is not made in an initial time of the order $T_1 Z$ before saturation is complete.

2.7. The Bloch susceptibilities

So far our approach to the nuclear magnetic resonance phenomenon has started from the magnetic properties of the nucleus and has led to certain macroscopic predictions. In this we have followed the arguments of Bloembergen, Purcell and Pound (1948). On the other hand, in developing his theory of nuclear induction, Bloch

(1946) argued in mainly macroscopic terms throughout. His more phenomenological approach is particularly well suited to the study of transient effects, while at the same time for systems in a steady state it is broadly consistent with the quantum treatment.† We reserve for the next section the discussion of nuclear induction, and in the present section we use Bloch's approach to derive expressions for the real and imaginary parts of the nuclear magnetic susceptibility.

Let us consider from a classical viewpoint a nucleus of magnetic moment μ in a magnetic field $\mathbf{H}$. The nuclear magnet experiences a couple ($\mu \wedge \mathbf{H}$), which must be equated to the rate of change of angular momentum. Since the angular momentum of the nucleus is μ/γ, we have therefore

$$\frac{1}{\gamma}\frac{d\mu}{dt} = \mu \wedge \mathbf{H}. \tag{2.43}$$

If $\mathbf{H}$ is constant, this equation of motion represents the precession of the vector μ about the vector $\mathbf{H}$ with angular frequency γH.

If now we consider an assembly of weakly interacting nuclear spins, then the magnetization vector $\mathcal{M}$ is the vector sum of the nuclear magnetic moments in unit volume. Summing (2.43) over unit volume we get

$$\frac{d\mathcal{M}}{dt} = \gamma \mathcal{M} \wedge \mathbf{H}. \tag{2.44}$$

This equation, which has been obtained from the classical equation (2.43), is nevertheless valid also in quantum mechanics. It is a well-known principle, fully exploited by Bloch, that the quantum-mechanical expectation value of any quantity follows in its time dependence just the classical equations of motion.

The steady-state solution of equation (2.44) for constant $\mathbf{H}$ is of the same form as that of (2.43) and represents the precession of the vector $\mathcal{M}$ about $\mathbf{H}$ with angular frequency γH. We shall, however, be concerned with the case where the three Cartesian components of $\mathbf{H}$ are not all constant; instead let us suppose their values are

$$H_x = H_1 \cos \omega t, \quad H_y = -H_1 \sin \omega t, \quad H_z = H_0. \tag{2.45}$$

† The assumptions implicit in Bloch's approach have been examined by Wangsness and Bloch (1953).

These components represent the actual situation in a nuclear magnetic resonance experiment. H_x and H_y together represent a magnetic field of amplitude H_1, rotating in a plane normal to the steady field H_0, in the same sense as the precession of a nucleus having a positive value of γ. (If γ should be negative the sign of H_y must be changed.)

Equation (2.44) is not quite complete in an actual nuclear spin system, since the right-hand term does not express all the mechanisms by which $\mathcal{M}$ may change. The component $\mathcal{M}_z$ represents the resultant magnetic moment per unit volume parallel to the steady field $\mathbf{H_0}$. In absence of the rotating magnetic field, and with the spin system and the lattice in thermal equilibrium, this component is given by

$$\mathcal{M}_0 = \chi_0 H_0. \tag{2.46}$$

If the spin system and lattice are not in thermal equilibrium, then in absence of a radiofrequency field $\mathcal{M}_z$ approaches $\mathcal{M}_0$ exponentially with the characteristic time T_1 as we saw in §2.2. In this case the z component of the equation of motion is

$$\frac{d\mathcal{M}_z}{dt} = \frac{\mathcal{M}_0 - \mathcal{M}_z}{T_1}. \tag{2.47}$$

This equation is just a rewriting of (2.8) and (2.11) remembering that $\mathcal{M}_z$ is proportional to the excess number of nuclei in the lower energy state. It is because T_1 determines the approach to equilibrium of the component of $\mathcal{M}$ parallel to $\mathbf{H_0}$ that Bloch has termed T_1 the *longitudinal* relaxation time as we mentioned in §2.2.

We see therefore that to complete the z component of the equation of motion in the general case we must add the right-hand side of (2.47) to the z component of (2.44).

The transverse components $\mathcal{M}_x$ and $\mathcal{M}_y$ represent the rotating component of the precessing magnetization vector $\mathcal{M}$. We saw in §2.3 that local irregularities of the magnetic field cause the individual precessing nuclei to get out of phase with each other in a time of the order of the spin-spin interaction time T_2. In absence of a radiofrequency magnetic field, any phase coherence of the nuclei would be destroyed in a time of the order T_2, thus bringing $\mathcal{M}_x$ and $\mathcal{M}_y$ to zero. Bloch therefore assumed for simplicity that the approach to zero is exponential, with characteristic time T_2, giving

$$\frac{d\mathcal{M}_x}{dt} = -\frac{\mathcal{M}_x}{T_2} \quad \text{and} \quad \frac{d\mathcal{M}_y}{dt} = -\frac{\mathcal{M}_y}{T_2}. \qquad (2.48)$$

The definition of T_2 in §2.3 has been so arranged, as we shall see later, that it is consistent with its use in (2.48). Because T_2 enters into these equations governing the time dependence of the transverse magnetization components $\mathcal{M}_x$ and $\mathcal{M}_y$, Bloch has used the name *transversal* relaxation time for T_2, as we mentioned in §2.3.

Thus to complete the x and y components of the equation of motion in the general case, we must add the right-hand sides of (2.48) to the x and y components of (2.44). On evaluating the vector product $(\mathcal{M} \wedge \mathbf{H})$, and substituting for the components of $\mathbf{H}$ from (2.45), the three component equations now become:

$$\left.\begin{aligned}
\frac{d\mathcal{M}_x}{dt} &= \gamma\left[\mathcal{M}_y H_0 + \mathcal{M}_z H_1 \sin \omega t\right] - \frac{\mathcal{M}_x}{T_2}, \\
\frac{d\mathcal{M}_y}{dt} &= \gamma\left[\mathcal{M}_z H_1 \cos \omega t - \mathcal{M}_x H_0\right] - \frac{\mathcal{M}_y}{T_2}, \\
\frac{d\mathcal{M}_z}{dt} &= \gamma\left[-\mathcal{M}_x H_1 \sin \omega t - \mathcal{M}_y H_1 \cos \omega t\right] + \frac{M_0 - \mathcal{M}_z}{T_1}.
\end{aligned}\right\} \quad (2.49)$$

The steady-state solution of these three differential equations, often referred to as the Bloch equations, is derived in Appendix 1. The result is:

$$\mathcal{M}_x = \tfrac{1}{2}\chi_0\omega_0 T_2\left[\frac{(2H_1 \cos \omega t)(\omega_0 - \omega)T_2 + 2H_1 \sin \omega t}{1 + (\omega_0 - \omega)^2 T_2^2 + \gamma^2 H_1^2 T_1 T_2}\right], \quad (2.50)$$

$$\mathcal{M}_y = \tfrac{1}{2}\chi_0\omega_0 T_2\left[\frac{2H_1 \cos \omega t - (2H_1 \sin \omega t)(\omega_0 - \omega)T_2}{1 + (\omega_0 - \omega)^2 T_2^2 + \gamma^2 H_1^2 T_1 T_2}\right], \quad (2.51)$$

$$\mathcal{M}_z = \chi_0 H_0\left[\frac{1 + (\omega_0 - \omega)^2 T_2^2}{1 + (\omega_0 - \omega)^2 T_2^2 + \gamma^2 H_1^2 T_1 T_2}\right]. \quad (2.52)$$

Since in an actual experiment the oscillating field is linearly polarized, of the form $H_x = 2H_1 \cos \omega t$, rather than circularly polarized, we find from (2.50) and (2.51) that the components χ' and χ'' of the nuclear magnetic susceptibility are

$$\chi' = \tfrac{1}{2}\chi_0\omega_0 T_2\left[\frac{(\omega_0 - \omega)T_2}{1 + (\omega_0 - \omega)^2 T_2^2 + \gamma^2 H_1^2 T_1 T_2}\right], \quad (2.53)$$

$$\chi'' = \tfrac{1}{2}\chi_0\omega_0 T_2\left[\frac{1}{1 + (\omega_0 - \omega)^2 T_2^2 + \gamma^2 H_1^2 T_1 T_2}\right]. \quad (2.54)$$

We recognize the term $\gamma^2 H_1^2 T_1 T_2$ in the denominators of equations (2.50)–(2.54) as being the dimensionless parameter determining the degree of saturation (§2.4). Let us therefore consider the form of the Bloch susceptibilities when H_1 is small enough to avoid appreciable saturation, namely when

$$\gamma^2 H_1^2 T_1 T_2 \ll 1. \tag{2.55}$$

In this situation we see from (2.52) that, as we should expect,

$$\mathscr{M}_z \cong \mathscr{M}_0 = \chi_0 H_0,$$

at all frequencies. The susceptibilities χ' and χ'' are now given by

$$\chi' = \tfrac{1}{2}\chi_0 \omega_0 T_2 \left[\frac{(\omega_0 - \omega) T_2}{1 + (\omega_0 - \omega)^2 T_2^2} \right], \tag{2.56}$$

$$\chi'' = \tfrac{1}{2}\chi_0 \omega_0 T_2 \left[\frac{1}{1 + (\omega_0 - \omega)^2 T_2^2} \right]. \tag{2.57}$$

These susceptibilities are plotted in fig. 4 as a function of the dimensionless product $(\omega_0 - \omega) T_2$. The curve for χ'' shows the resonant character of the absorption, while the curve for χ' shows the dispersion which accompanies the absorption.

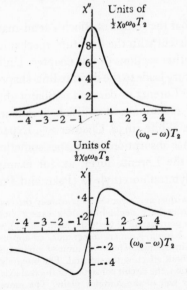

Fig. 4. The Bloch nuclear susceptibilities plotted against the dimensionless product $(\omega_0 - \omega) T_2$. The graphs apply to conditions of negligible saturation. (Pake, 1950.)

Expression (2.57) should, of course, be consistent with the more general expression for χ'', (2.40), which was derived earlier. Since we are treating the case where there is no saturation, $Z=1$, and (2.40) gives

$$\chi'' = \tfrac{1}{2}\pi\chi_0\nu_0 g(\nu).\tag{2.58}$$

If (2.57) and (2.58) are to be consistent, then the shape function $g(\nu)$ must be given by

$$g(\nu) = \frac{2T_2}{1 + 4\pi^2(\nu_0 - \nu)^2 T_2^2},\tag{2.59}$$

and the right-hand side of this equation must be a normalized function of the frequency ν. A straightforward integration shows that this is so. Furthermore, we see that $g(\nu)_{max}$ is equal to $2T_2$, as it should be by (2.16). In fact, the definition of T_2 expressed by equation (2.16) was so designed as to be consistent with Bloch's usage. By forming the product $Zg(\nu)$, using (2.24) and (2.59), we see that the more general expressions (2.40) and (2.54) for $Z\neq 1$ are also in agreement. We may further note that the susceptibilities χ' and χ'' given by (2.56) and (2.57) satisfy the Kronig–Kramers relations (2.41).†

Thus we see that the results of Bloch's semi-macroscopic theory are broadly consistent with the quantum-mechanical approach developed in the earlier sections of this chapter. Unlike the quantum approach, his theory leads to the definite line-shape given by (2.59). This shape is the Lorentz or damped oscillator shape encountered in the theory of radiation and collision-broadened spectral lines (see, for example, White, 1934, Chapter 21). Experimental nuclear magnetic resonance absorption line-shapes sometimes approximate quite closely to the Lorentz form, as, for example, for the ^{19}F resonance in polytetrafluorethylene (Pake and Purcell, 1948). In

† In order to show this, one assumes, as is usually the case, that $\omega_0 T_2 \gg 1$, so that the susceptibilities have negligible values except close to the resonant frequency. It is then permissible to write the factor $(\nu' + \nu)$, which occurs in the denominator of the Kronig–Kramers integrands, as approximately $2\nu'$ or 2ν, thus simplifying the integrands. The lower limit of integration may be extended to minus infinity without affecting the integral. The integrals are then evaluated by a contour integration, the circuit consisting of the real axis and a semicircle at infinity in the upper half of the complex plane. The more general equations (2.53) and (2.54) do not satisfy the Kronig–Kramers relations, since, as was pointed out in the last section, these relations apply only to systems which are not being saturated.

other cases, where the approximation is not quite so good, the Lorentz shape is nevertheless still a useful guide to behaviour. However, in the case of solids, as will be seen in Chapter 6, the actual shape is often markedly different from the Lorentz shape, and may, indeed, have more than one maximum. This difference is emphasized by one curious feature of the Lorentz line-shape which concerns its mean square width, or second moment,

$$\int_0^\infty (\nu_0 - \nu)^2 g(\nu) d\nu. \qquad (2.60)$$

By a straightforward integration one finds that for this line-shape the mean square width is infinite, due to the very slow convergence of the 'wings' of the curve; the experimental values of mean square width are finite.

2.8. Nuclear induction

It may be seen from (2.50) and (2.51) that in the steady state the magnetization components $\mathscr{M}_x$ and $\mathscr{M}_y$ constitute a magnetic moment of constant amplitude rotating in the xy plane with angular frequency ω. Thus along any line in this plane the component of the magnetic moment is oscillating with angular frequency ω. If a cylindrical coil of Λ area-turns surrounds the assembly of nuclear spins, with its axis in the xy plane, the oscillating flux of magnetic induction causes an electromotive force to be induced in the coil. The amplitude of this e.m.f. will clearly not depend on the orientation of the axis of the coil in the xy plane; the orientation merely determines the phase. For simplicity let us place the coil axis along the y axis; it is thus at right angles to the steady magnetic field H_0 along the z axis, and also at right angles to the oscillating field $2H_1 \cos \omega t$ along the x axis. The magnetic induction accompanying the magnetic moment $\mathscr{M}_y$ is

$$B_y = 4\pi \mathscr{M}_y = 8\pi H_1(\chi' \sin \omega t - \chi'' \cos \omega t). \qquad (2.61)$$

The flux of this magnetic induction through the coil is therefore

$$\Phi_y = \zeta \Lambda B_y, \qquad (2.62)$$

where ζ is the filling factor denoting the proportion of the effective coil area which is occupied by the spin system. Using Faraday's law of magnetic induction, the induced e.m.f. is found from (2.62) and (2.61) to be

$$\mathscr{E} = -\frac{d\Phi_y}{dt} = -\zeta \Lambda 8\pi H_1 \omega(\chi' \cos \omega t + \chi'' \sin \omega t). \quad (2.63)$$

The amplitude of the in-phase and quadrature components are thus proportional to χ' and χ'' respectively, and may be amplified up and observed; the details of the experimental arrangement are discussed in §3.6. Since the e.m.f. is obtained directly by electromagnetic induction from the precessing resultant of the nuclear magnetic moments, Bloch therefore aptly named the phenomenon *nuclear induction*.

The situation we have just discussed thus calls for two coils to be wound round the material containing the nuclear spins: one coil, called the transmitter, has its axis along the x axis, and when an oscillating current is passed through it, provides the radiofrequency field $2H_1 \cos \omega t$; the other coil, called the receiver, is the one we have just described, having area-turns Λ, with its axis along the y axis. If the geometrical arrangement is perfect, this disposition of the two coils has the advantage that the receiver coil picks up no radiofrequency e.m.f. by direct induction from the transmitter. In principle, however, as we have just said, the axis of the receiving coil could lie anywhere in the xy plane, without affecting the nuclear induction signal (apart, of course, from its phase). In particular, the transmitting and receiving coil could both lie along the same axis, in our case the x axis. We may then go on to argue that they could both be one and the same coil. The experimental arrangement now becomes just that for nuclear magnetic resonance absorption outlined in §2.6. Moreover, as we will now show, we are in fact carrying out precisely the same experiment.

Let us calculate the change of potential difference across the coil supplying the radiofrequency field from the viewpoint of nuclear magnetic resonance absorption. We will suppose that the radiofrequency field is generated by loosely coupling the coil to a radiofrequency source, which may therefore be treated as a constant current source. In order to simplify this calculation we will suppose that the coil has no tuning condenser in parallel with it. In practice such a condenser is usually present, but this does not affect the conclusions to be drawn here. Since the inductance of a coil is proportional to the magnetic flux linking it, the change of inductance of the coil near resonance is

$$\Delta \mathscr{L} = 4\pi(\chi' - i\chi'')\zeta\mathscr{L}, \tag{2.64}$$

where, as before, ζ is the filling factor. The change of impedance is therefore $i\omega\Delta\mathscr{L}$, and as a result there is a change of potential difference across the coil given by

$$\Delta \mathscr{V} = 4\pi\omega\zeta\mathscr{L}\mathscr{I}_0(-\chi' \sin \omega t + \chi'' \cos \omega t), \tag{2.65}$$

where $\mathscr{I}_0 \cos \omega t$ is the radiofrequency current. From the definition of inductance we have

$$\mathscr{L}\mathscr{I}_0 = 2H_1\Lambda, \tag{2.66}$$

where Λ, as before, is the area-turns of the coil, which is assumed for simplicity to be identical with the receiving coil previously considered. Combining (2.65) and (2.66) we find that the potential difference is

$$\Delta \mathscr{V} = \zeta\Lambda 8\pi H_1\omega(-\chi' \sin \omega t + \chi'' \cos \omega t). \tag{2.67}$$

This potential difference $\Delta \mathscr{V}$ is identical with nuclear induction e.m.f. $\mathscr{E}$ given by (2.63) apart from a phase difference of $\frac{1}{2}\pi$ which is caused by placing the nuclear induction receiving coil at right angles to the transmitting coil. (Since the receiving coil is considered to be untuned for the purpose of this calculation, the potential across the coil produced by the nuclear induction e.m.f. is just equal to that e.m.f. In practice the coil is usually tuned; this has the effect from the point of view of both methods of increasing the potential difference by the quality factor $\mathscr{Q}$ of the circuit.)

This equality of voltages is no coincidence, since it merely expresses the fact that absorption in any system at a given frequency may be simulated by replacing the absorber with an oscillator of the same frequency but of opposite phase, whilst dispersion may be simulated by a similar oscillator with phase in quadrature. However, an essential feature of Bloch's argument (Bloch 1951 a) is that such oscillators are not a fiction, but are actually present in the form of the forced precession of the resultant magnetic moment about $\mathbf{H}_0$. Moreover, Bloch points out that in certain experiments, which will be discussed in §5.5, one may continue to observe nuclear induction of the freely precessing magnetic moment after the radiofrequency field has been switched off. For these reasons Bloch considers the term *nuclear induction* to be a more appropriate description of the whole class of phenomena than *nuclear magnetic resonance*. Both terms are used in the literature. However, the tendency is to restrict the term nuclear induction to cases where the experimental arrangement is similar to that proposed by Bloch, with separate transmitting and receiving coils or where the behaviour is more readily comprehended in terms of his semi-macroscopic theory. We shall follow this tendency in this book.

The general conditions for observing nuclear induction are similar to those enumerated in §2.6 for nuclear magnetic resonance absorption, with the difference that the signal is picked up in the second (receiving) coil, rather than by its reaction on the driving circuit. Since the two experiments are identical in principle as we have just shown, the choice of experimental arrangement is mainly determined by convenience for the particular conditions of observation. The first successful nuclear induction experiments (Bloch, Hansen and Packard, 1946 *a*, *b*) were concerned with transient rather than steady-state conditions such as we have been discussing here. We shall, however, defer a discussion of transient effects to §5.5.

CHAPTER 3

EXPERIMENTAL METHODS

3.1. Rollin's arrangement for detecting nuclear magnetic resonance absorption

As we pointed out in the previous chapter, the essential feature of nuclear magnetic resonance absorption experiments, which distinguishes them from nuclear induction experiments, is that the resonance effects are detected by their reaction on the circuit supplying the radiofrequency field. The arrangement used by Rollin (1946, 1949) in his pioneer experiments at low temperatures is probably the simplest. An essentially similar apparatus was later described by Gabillard (1951 a).

Suppose we wish to detect the resonance of protons in water at room temperature. About 1 cm.³ of water is placed in a sealed glass tube, which, as shown in fig. 5, is put within the coil of a tuned

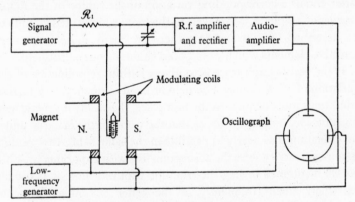

Fig. 5. Schematic diagram of Rollin's arrangement for observing nuclear magnetic resonance.

circuit and held between the poles of a magnet supplying the steady field H_0. Radiofrequency current is supplied to the coil through a high impedance $\mathscr{R}_1$ from a signal generator, which thus acts effectively as a constant-current source. At resonance the radiofrequency power absorbed by the protons causes a decrease in the $\mathscr{Q}$ (quality factor) and in the parallel impedance of the tuned circuit, which in turn produces a drop in radiofrequency voltage across the circuit. This radiofrequency voltage is amplified and rectified. In order to display the resonance absorption clearly, it is common practice to modulate the 'steady' magnetic field H_0 at a low audiofrequency, often about 25 c./s., with an amplitude of a few gauss, by means of auxiliary magnet coils. When the mean value of the steady field is close to the resonant value, the field is swept twice through the resonant condition each cycle, producing an audio-amplitude modulation of the radiofrequency carrier. After rectification, the audio signal is further amplified and it is then displayed on a cathode-ray oscillograph, the timebase of which runs in synchronism with the modulation. The oscillograph trace then gives a visible indication of the absorption line, as exemplified in fig. 6.

It will be noticed that we have suggested that the field be modulated while maintaining a constant radiofrequency. We could equally well have frequency-modulated the radiofrequency and held the magnetic field constant. It is, however, generally simpler to vary the magnetic field than to vary the frequency, since the latter entails a corresponding variation of the tuning of the radiofrequency amplifier. Moreover, in the more complicated circuits to be described in §3.2 and §3.3 there are many more frequency-sensitive elements which would need simultaneous adjustment.

If the oscillograph trace is to give a faithful reproduction of the variation of χ'' with magnetic field at fixed frequency, two further conditions must be met. In the first place, the steady magnetic field must be very homogeneous, so that the whole of the material under investigation is as nearly as possible in the same field. Any residual inhomogeneity endows the absorption line with a spurious breadth. As we mentioned in §2.3 the true line width for liquids is usually extremely small (frequently of the order of milligauss); consequently, the observed line width for liquids is often determined by the field inhomogeneity and not by χ''.

The second requirement which must be met if the trace is to reproduce the variation of χ'' with field, is that conditions shall approximate to those of the steady-state absorption to which χ'' refers. As we shall see in §5.5, where transient effects are discussed in more detail, steady-state conditions are approximated as follows:

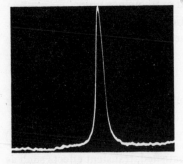

Fig. 6. Nuclear magnetic resonance absorption curve for the protons in a solution of ferric nitrate. An oscillogram taken from the early work of Bloembergen, Purcell and Pound (1948).

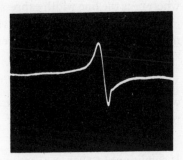

Fig. 8. Nuclear magnetic resonance dispersion curve for the protons in a solution of ferric nitrate. An oscillogram taken from the early work of Bloembergen, Purcell and Pound (1948).

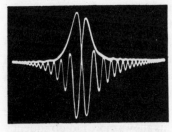

Fig. 7. An example of the transient effect ('wiggles'), which is discussed in more detail in § 5.5.1, here observed in paraffin oil (courtesy F. A. Rushworth). The decaying oscillations always appear after passage through the resonance condition. Since the resonance condition is traversed from left to right and from right to left once each per modulation cycle, the oscillogram shows two superimposed sets of decaying oscillations.

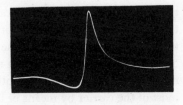

Fig. 9. A mixed absorption and dispersion curve, which results from the use of a radiofrequency bridge which is not correctly balanced (see § 3.2).

(a) If the frequency is varied with fixed field, then on sweeping through resonance, the rate of change of frequency ν must satisfy the requirement

$$\frac{d\nu}{dt} \ll (\delta\nu)^2, \tag{3.1}$$

where $\delta\nu$ is the line width expressed in frequency units.

(b) If the steady field H_0 is varied at fixed frequency, then we must have the equivalent requirement

$$\frac{dH_0}{dt} \ll |\gamma| (\delta H_0)^2, \qquad (3.2)$$

where δH_0 is the line width expressed in field units, which is related to $\delta\nu$ by

$$2\pi\delta\nu = |\gamma| \delta H_0. \qquad (3.3)$$

If conditions (3.1) or (3.2) are not satisfied, transient effects such as those shown in fig. 7 may be obtained, the trace no longer reproducing the steady-state absorption curve.

The radiofrequency voltage across the tuned circuit usually lies within the range 10^{-3} to 10 V. With Rollin's simple circuit there cannot therefore be a great deal of linear amplification before detection, and this makes it difficult to get a good noise factor for the amplifier. A good noise factor is frequently of importance, since the depth of amplitude modulation, which represents the absorption signal, is very small, and the limit of sensitivity for observation of signals is set by the random noise background. For this reason many workers enhance the depth of modulation by balancing out most of the carrier by means of bridge arrangements which will be described in the next two sections. This allows more radiofrequency amplification before detection and thus a better noise factor may be obtained. A further advantage accrues from balancing out most of the carrier voltage, namely, that extraneous noise introduced from the signal generator as a modulation of the carrier is balanced out to the same degree. Against these superior features possessed by the bridge circuit, Rollin's circuit possesses the advantages (a) of simplicity, and (b) of providing a direct measure of χ'', while the bridge method, as we shall see, gives a mixture of χ'' and χ' unless carefully adjusted.

However, before proceeding to a discussion of these bridge circuits we will first calculate the depth of modulation produced by the absorption, and show that it is indeed small. From equation (2.64) we see that the effect of nuclear magnetic resonance on the radiofrequency coil is to increase its inductance $\mathscr{L}$ by a fraction $4\pi\chi'\zeta$, and to increase its effective series resistance $\mathscr{R}_0$ by a fraction

$4\pi\chi''\zeta\mathscr{Q}$, where $\mathscr{Q}$, the quality factor of the coil, is given by $\omega\mathscr{L}/\mathscr{R}_0$. An elementary analysis† of a parallel circuit of high $\mathscr{Q}$, tuned to the radiofrequency in use, shows that to a first order a small fractional increase in the series resistance of the coil results in an equal fractional decrease in the magnitude of the parallel impedance of the circuit, the phase angle of that impedance remaining unchanged. One further finds to a first order that a small fractional increase in the inductance of the coil results in a change of phase angle of the parallel impedance equal to $\mathscr{Q}$ times that fraction, while the magnitude of the parallel impedance remains unchanged. Thus in our problem the parallel impedance is reduced in magnitude by a fraction $4\pi\chi''\zeta\mathscr{Q}$. If the tuned circuit is fed from a constant-current source, the value of this fraction, when χ'' has its greatest value, is evidently the degree of amplitude modulation‡ of the carrier resulting from the absorption. Similarly, there is a change in phase angle given by $4\pi\chi'\zeta\mathscr{Q}$, which results in a phase modulation of the carrier. The normal type of rectifying detector envisaged in fig. 5 does not, however, respond to phase modulation, but gives an output proportional to the carrier envelope, namely, proportional to χ''.

In absence of saturation the maximum value of χ'' is found from Bloch's formulation (2.57) to be $\frac{1}{2}\chi_0\omega_0 T_2$. The degree of amplitude modulation is therefore $2\pi\chi_0\omega_0 T_2\zeta\mathscr{Q}$. The static susceptibility χ_0 is given by (2.33), and for protons in water at room temperature is found to be 3×10^{-10} erg gauss^{-2} cm.$^{-3}$. Taking $\zeta=1$, $\mathscr{Q}=100$, $\omega_0/2\pi=21\cdot3$ Mc./s. (corresponding to $H_0=5000$ gauss), and

---

† The admittance $1/\mathscr{Z}$ at angular frequency ω of a parallel combination of a condenser of capacity C and a coil of inductance $\mathscr{L}$ with small series resistance $\mathscr{R}_0$, is $[i\omega C + (\mathscr{R}_0 + i\omega\mathscr{L})^{-1}]$. The impedance $\mathscr{Z}$ is therefore

$$(\mathscr{R}_0 + i\omega\mathscr{L})/(1 - \omega^2\mathscr{L}C + i\omega C\mathscr{R}_0),$$

and when the circuit is tuned to resonance is

$$\mathscr{Z}_0 = (\mathscr{R}_0 + i\omega\mathscr{L})/i\omega C\mathscr{R}_0 \simeq \omega^2\mathscr{L}^2/\mathscr{R}_0$$

(purely resistive), since $\omega\mathscr{L}/\mathscr{R}_0 = \mathscr{Q} \gg 1$. In this condition differentiation gives $\partial\mathscr{Z}/\partial\mathscr{R}_0 = -\mathscr{Z}_0/\mathscr{R}_0$, which is real; thus a fractional increase $\delta\mathscr{R}_0/\mathscr{R}_0$ in coil resistance results in a purely resistive equal fractional reduction in $\mathscr{Z}$, since $\delta\mathscr{Z}/\mathscr{Z}_0 = -\delta\mathscr{R}_0/\mathscr{R}_0$. Similarly, differentiation gives $\partial\mathscr{Z}/\partial\mathscr{L} = -i2\mathscr{Z}_0/\mathscr{L}$, which is purely imaginary; thus a fractional increase $\delta\mathscr{L}/\mathscr{L}$ in coil inductance results in a change of phase angle $|\delta\mathscr{Z}|/\mathscr{Z}_0 = \mathscr{Q}\,\delta\mathscr{L}/\mathscr{L}$.

‡ The degree of amplitude modulation (or modulation factor) may be defined as the maximum fractional reduction in amplitude of the carrier wave.

$T_2 = 3 \times 10^{-4}$ sec. (equivalent to a typical inhomogeneity line width of about 0·2 gauss), the degree of amplitude modulation is found to be about 0·01. Such a depth of modulation presents little difficulty, and in fact the absorption signal in water is strong and is readily detected. In general, however, the situation is much less favourable. If the material investigated is a solid, the line width may be about 10 gauss wide (see §2.3), causing T_2 and the degree of modulation to be some 50 times smaller. The degree of modulation is also proportional to χ_0 and therefore to $\mu^2(I+1)/I$; for very few nuclei is this quantity as large as it is for the proton, while it may well be 10 times less. The static susceptibility is also proportional to N, the number of nuclei per cm.[3]. This number might be as much as 100 times smaller than for protons in water if the material is gaseous, or if the abundance of the particular isotope concerned is small, or if the chemical proportion of the particular element is small. Again, the degree of amplitude modulation is proportional to the resonance frequency $\omega_0/2\pi$ which, in a given steady field, is proportional to the nuclear gyromagnetic ratio. With the exception of the triton, the proton has the highest gyromagnetic ratio of all nuclear species, and another factor of 10 may well be lost here. The degree of amplitude modulation may therefore in less favourable cases be as low as 10^{-5} or less.

3.2. The bridge arrangement of Bloembergen, Purcell and Pound

In the first successful nuclear magnetic resonance absorption experiment, Purcell, Torrey and Pound (1946) used a bridge circuit, the details of which have been given by Bloembergen (1948). We shall not, however, describe this bridge here, since it proved to be inconvenient in general use, and was replaced by the arrangement of Bloembergen, Purcell and Pound (1948), shown in fig. 10.

Two almost identical tuned circuits are fed in parallel with radiofrequency power through properly terminated cables and small coupling condensers C_1 and C_2. The coil of one circuit contains the material under investigation and is placed between the poles of the magnet. By placing an additional electrical half-wavelength of cable in one circuit, say at AB, the output of the two circuits may be subtracted at A and fed into an amplifier. If the two circuits were identical, and if AB were an exact half-wave-length of loss-free cable, the signal from the dummy circuit would exactly cancel that from the circuit containing the experimental

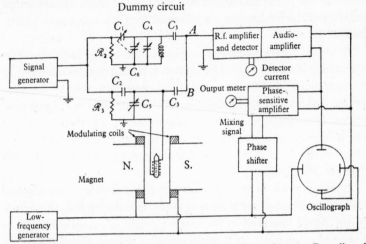

Fig. 10. Schematic diagram of the apparatus of Bloembergen, Purcell and Pound (1948). In a typical arrangement for work at 30 Mc./s. the circuit components were: $\mathscr{R}_2 = \mathscr{R}_3 = 50$ ohms (the characteristic impedance of the coaxial cable used); $C_1 = C_6 = 5$ pF. maximum; $C_2 = 3$ pF.; $C_3 = 4$ pF.; $C_4 = C_5 = 60$ pF. A typical coil consists of 12 turns 18 s.w.g. copper wire wound to give a coil 1·5 cm. long and 0·7 cm. inside diameter.

material, when the nuclei are not at resonance. The half-wave-length of cable could of course be placed in either arm of the bridge, and could precede the tuned circuit instead of following it. The small output coupling condensers C_3 are chosen to make the impedance seen when looking back from the amplifier approximately equal to the value for which the amplifier noise factor is best. The details of a procedure using a noise diode for finding the optimum value of these condensers has been described by Bloembergen (1948). The main part of the amplifier is often a communication receiver; a preamplifier is then desirable to improve the noise factor at the particular frequency of operation.

The bridge is balanced in phase and amplitude by adjustment of the tuning condenser C_4 and the coupling condenser C_1. The latter adjustment not only affects the amplitude balance, but unfortunately also affects the phase balance, since the coupling condenser C_1 is effectively in parallel with the tuned circuit, the terminating resistance $\mathscr{R}_2$ being small. The coupling condenser C_1 is therefore ganged in opposite sense with a similar condenser C_6 so that the total tuning capacity is unaltered during the change in coupling. In this way the phase and amplitude controls are made almost orthogonal, and the bridge may be balanced by a few successive adjustments.

It will be recalled that the purpose of the bridge is (i) to reduce the carrier level to a value low enough to permit considerable amplification before rectification, and (ii) to cancel out most of the noise arising from spurious modulation of the output of the signal generator. In order to achieve these requirements, a perfect balance of the bridge is not required; it usually suffices that the carrier level after combination at A be reduced to about 1% of its level in either arm (40 db. balance).

It is important to be able to control the character of the residual unbalance, since as we shall now show, this determines the nature of the signal observed as the magnetic field is varied through the region of resonance. If $\mathscr{V}_0$ is the complex voltage across the coil in the magnet gap when far from the nuclear resonance condition, then at or near resonance the voltage becomes

$$\mathscr{V}_0[1 - 4\pi\zeta\mathscr{Q}(\chi'' + i\chi')], \tag{3.4}$$

since, as mentioned in the previous section, the parallel impedance of the tuned circuit is reduced by a small fraction $4\pi\chi''\zeta\mathscr{Q}$ and the phase is changed by a small angle $4\pi\chi'\zeta\mathscr{Q}$. This voltage is added at A to the almost equal and almost oppositely phased voltage $-\mathscr{V}_1$ from the dummy circuit. The vector resultant $\mathscr{V}_R$ is therefore given by

$$\begin{aligned}\mathscr{V}_R &= (\mathscr{V}_0 - \mathscr{V}_1) - 4\pi\zeta\mathscr{Q}\mathscr{V}_0(\chi'' + i\chi') \\ &= |\mathscr{V}_0 - \mathscr{V}_1|\, e^{i\theta} - 4\pi\zeta\mathscr{Q}\,|\mathscr{V}_0|\, e^{i(\theta+\phi)}(\chi'' + i\chi'), \end{aligned} \tag{3.5}$$

where θ is the phase angle of $(\mathscr{V}_0 - \mathscr{V}_1)$ and $(\theta + \phi)$ is the phase angle of $\mathscr{V}_0$. Then

$$\mathscr{V}_R e^{-i\theta} = |\mathscr{V}_0 - \mathscr{V}_1| - 4\pi\zeta\mathscr{Q}\,|\mathscr{V}_0|\, e^{i\phi}(\chi'' + i\chi'). \tag{3.6}$$

Let us now arrange that the voltage $|\mathscr{V}_0 - \mathscr{V}_1|$, though less than either $|\mathscr{V}_0|$ or $|\mathscr{V}_1|$, is still large compared with the amplitude of the final term in (3.6). This means that the bridge must not be balanced too closely. To a good approximation $|\mathscr{V}_R|$ is then given by the sum of the comparatively large first term on the right in (3.6) and the component of the small second term along the direction of the first term. Since $|\mathcal{V}_0 - \mathcal{V}_1|$ is real, we therefore take the real part of the second term, giving

$$|\mathcal{V}_R| = |\mathcal{V}_0 - \mathcal{V}_1| + 4\pi\zeta\mathcal{Q}|\mathcal{V}_0|(-\chi'' \cos\phi + \chi' \sin\phi). \qquad (3.7)$$

As the modulated 'steady' field sweeps through the region of resonance a curve is traced out of the form

$$(-\chi'' \cos\phi + \chi' \sin\phi), \qquad (3.8)$$

since all the other terms in (3.7) are constant. Let us consider two important special cases.

(a) $\phi = 0$ or π. In this condition the oscillograph plots the absorption curve χ'', and requires $\mathcal{V}_0$ and $\mathcal{V}_1$ to have exactly the same phase. There is thus an unbalance of amplitude only.

(b) $\phi = \frac{1}{2}\pi$ or $\frac{3}{2}\pi$. In this condition the oscillograph plots the dispersion curve. Since we require a phase difference of $\frac{1}{2}\pi$ between $\mathcal{V}_0$ and $(\mathcal{V}_0 - \mathcal{V}_1)$, it follows that $\mathcal{V}_0$ and $\mathcal{V}_1$ must be equal in amplitude, leaving only an unbalance in phase.

While the absorption curve usually has a symmetrical shape (see, for example, fig. 4), and the dispersion curve is usually antisymmetrical, intermediate values of ϕ give rise to an unsymmetrical mixture of the two curves. An absorption curve, dispersion curve and a mixed curve are shown in figs. 6, 8 and 9. The condition of the bridge which determines the nature of the curve displayed is usually maintained by observation of the detector current, which gives an indication of the carrier level. As an alternative indicator of carrier level H. L. Anderson (1949) has used a panoramic amplifier.

The strength of the signal compared with the inherent noise background is very often insufficient to allow display on an oscillograph. Such a display calls for an amplifier bandwidth of the order of 10^3 c./s. A considerable improvement in signal-to-noise ratio is achieved by drastically reducing the bandwidth, thus accepting noise from a much narrower region of the noise spectrum. It is then necessary to display the output signal on a meter, often a recording meter. The time constant of the meter and its associated circuit usually determines the bandwidth of the system. For this purpose a phase-sensitive or 'lock-in' amplifier is used and is indicated in the scheme shown in fig. 10.

An elementary form of phase-sensitive amplifier is shown schematically in fig. 11 a. The signal, which consists of a waveform having the periodicity of the audio-field modulation is fed at A

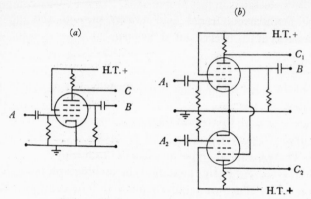

Fig. 11. Phase-sensitive detector circuits.

together with noise, on to the control grid of a pentode valve. A sinusoidal signal of the modulation frequency, derived from the audio-modulation source, is fed on to the suppressor grid through B. The pentode thus acts as a mixer, excursions of the voltage at C being proportional to the product of the voltages fed in at A and B.

Suppose the voltage $f(t)$ fed in at A is periodic, though not necessarily simple harmonic, and has the angular frequency ω_m of the audio modulation of the steady field, so that we may express it as the Fourier expansion

$$f(t) = \sum_{p=1}^{\infty} b_p \sin (p\omega_m t + \epsilon_p), \qquad (3.9)$$

where b_p and ϵ_p are the amplitude and phase of the pth harmonic or Fourier component. Then the voltage $\mathscr{V}_c$ at C is proportional to

$$f(t) \sin \omega_m t = \sum_{1}^{\infty} b_p \sin \omega_m t \sin (p\omega_m t + \epsilon_p). \qquad (3.10)$$

The mean voltage $\overline{\mathscr{V}_c}$ at C is therefore proportional to

$$\sum_{1}^{\infty} b_p \int_{0}^{2\pi/\omega_m} \sin \omega_m t \sin (p\omega_m t + \epsilon_p) dt. \qquad (3.11)$$

Expanding the second factor of the integrand and eliminating all except one of the terms in the summation on account of the orthogonal property of sines and cosines, we get

$$\mathscr{V}_c \propto b_1 \cos \epsilon_1. \qquad (3.12)$$

The mean voltage is in fact proportional to the Fourier component b_1 of the fundamental frequency. If the voltage from C is applied to a high-impedance backed-off voltmeter, the reading will give a measure of b_1. It will be noticed that the reading is phase-sensitive, being proportional to $\cos \epsilon_1$ and is therefore greatest when ϵ_1, the phase angle between the fundamental frequency voltages at A and B is o or π, so that $|\cos \epsilon_1|$ is unity. The mixing signal must therefore be arranged to have the correct phase.

The device also extracts the Fourier components of the noise at or near the frequency $\omega_m/2\pi$; if the meter and its associated circuit have a time constant of t_c, noise components in a bandwidth of the order $1/t_c$ around $\omega_m/2\pi$ are registered as fluctuations on the meter. The time constant t_c is often about 1 sec., but may if necessary be made much longer, say 100 sec., so that the bandwidth is cut down to values of the order of 1 c./s. or 10^{-2} c./s. respectively.

The simple circuit shown in fig. 11a is not satisfactory in practice, since changes in supply voltage and changes in gain of the pentode caused by changes of supply voltage or of temperature produce spurious fluctuations at C even in absence of noise fed in at A. A first-order compensation for these variations is provided by the push-pull arrangement of fig. 11b. This is the arrangement used in the circuit developed by Dicke (1946), which has been much employed in this field of work; details of the circuit have been given by Bloembergen (1948). A balanced signal is applied at A_1 and A_2 to the control grids of identical pentode valves, while the reference signal is applied at B to both suppressor grids. A high-impedance voltmeter examines the average difference in potential on the anodes at C_1 and C_2. Improved circuits have been developed by Schuster (1951) and Cox (1953).

It will be noticed that if the input signal is not modulated by a simple harmonic term $\sin \omega_m t$, but by a more general function of angular frequency ω_m containing higher order Fourier components, the circuit then responds to harmonics of ω_m in the input signal and noise spectrum, in addition to the fundamental. This is not important, however, for in practice the harmonics are filtered off in earlier audio-amplifier stages, and in any case, as we shall now see, the input signal usually has little harmonic content even before these stages are reached.

When the nuclear magnetic resonance absorption or dispersion line is displayed on an oscillograph, it is usual to modulate the steady field with an amplitude several times the width of the absorption line, so that the resonance is swept right through and back again in each modulation cycle. On the other hand, when the phase-sensitive amplifier is used, the modulation amplitude is usually cut down to a fraction of the line width. The output signal is then sinusoidal with an amplitude proportional to the first derivative of the line shape; strictly this is only true for a very small amplitude of modulation such that the portion of the line shape which is traversed can be treated as linear. Provided the modulation amplitude is not more than about one-sixth of the interval between an adjacent maximum and minimum of the derivative, the curve obtained is a reasonably faithful reproduction of the derivative (Andrew, 1953).

As an alternative to the phase-sensitive amplifier for reducing the bandwidth and thus improving the signal-to-noise ratio Schneider (1948) has suggested the use of an oscillating audio detector. Other workers have obtained an improvement in signal-to-noise ratio by photographic integration (Bloch and Garber, 1949; Ross and Johnson, 1951), and by integration on a magnetic sound recorder (Suryan, 1950).

3.3. Other bridge circuits

The bridge circuit described in the previous section is shown schematically in fig. 12 a. As an alternative to the use of a half-wave line for subtracting the outputs of the specimen and dummy circuits a centre-tapped transformer may be used as shown in fig. 12 b (Pake, 1948); the two halves of one winding have opposite sense, so that the two signals are coupled to the output circuit in opposite phase. The transformer could equally well be placed on the input side of the circuit, as shown in fig. 12 c, so that the two circuits are supplied with radiofrequency current of opposite phase (Torrey,

1949). This removal of the half-wavelength of cable renders the circuit more symmetrical so that it is troubled less by any spurious

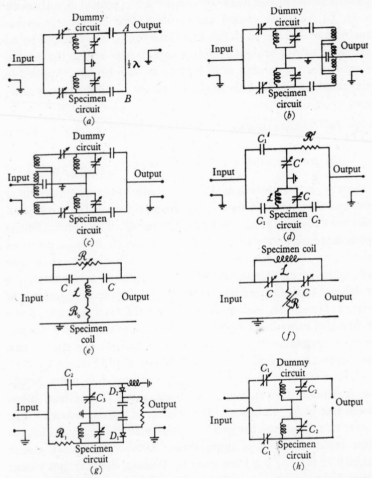

Fig. 12. Radiofrequency bridge circuits. (a) Circuit of Bloembergen, Purcell and Pound (1948) using a half-wave line AB. (b) Circuit using a centre-tapped transformer in place of the half-wave line (Pake, 1948). (c) Circuit with the centre-tapped transformer on the input side of the bridge (Torrey, 1949). (d) Asymmetrical bridge used by H. L. Anderson (1949). For work at 20 Mc./s. the circuit values used were $C_1 = 1$ pF., $C_2 = 3$ pF., $C = 65$ pF., $C' = 39$ pF., $C_1' = 6$ pF., $\mathscr{R}' = 100$ ohm, coil $\mathscr{L} = 0.89$ μH. ($\mathscr{Q} = 240$). (e) Bridged-T circuit (Grivet, Soutif and Buyle, 1949). (f) Alternative bridged-T circuit (Waring, Spencer and Custer, 1952). (g) Circuit of Thomas and Huntoon (1949) with immediate rectification. (h) Wheatstone bridge arrangement for low frequencies (Brown, 1950).

frequency modulation which may be present in the signal generator output. In addition, the removal of the cable usually makes the balance of the bridge more stable against mechanical disturbances.

H. L. Anderson (1949) has employed the deliberately asymmetrical bridge shown in fig. 12d. This bridge was inspired by an article on twin-T bridges by Tuttle (1940), who derives the general formula for the conditions of balance of such bridges. Applied to the circuit of fig. 12d, these conditions for a radiofrequency $\omega/2\pi$ are:

(a) resistive balance: $\omega^2 C_1 C_2 \left(1 + \dfrac{C'}{C_1'} \right) \mathscr{R}' \mathscr{R}_p = 1,$ (3.13)

(b) reactive balance: $C + C_1 + C_2 \left(1 + \dfrac{C_1}{C_1'} \right) = \dfrac{1}{\omega^2 \mathscr{L}},$ (3.14)

where $\mathscr{R}_p$ is the equivalent shunt resistance of the specimen coil at the radiofrequency in use. It will be seen that (3.13) involves C' but not C, while (3.14) involves C but not C'. The two conditions of balance may thus be reached independently by adjustment of the appropriate condenser. This bridge therefore has the advantage of improved orthogonality. Since the bridge is asymmetric, the balance is more frequency-dependent than for the arrangements of figs. 12b and 12c, and more care must therefore be paid to the removal of unwanted frequency modulation of the signal generator output. Attention was independently drawn to the use of this bridge by Grivet, Soutif and Gabillard (1951) and Soutif (1951).

The French workers had earlier suggested two other bridge arrangements (Grivet, Soutif and Buyle, 1949). One of these differs from that of fig. 12d only in the absence of the condenser C. The other circuit was the bridged-T arrangement of fig. 12e, which has low input and output impedances. Another bridged-T circuit shown in fig. 12f has been used by Waring, Spencer and Custer (1952). In comparison with the circuit of fig. 12d this circuit has the disadvantage that neither end of the specimen coil nor those of the two variable condensers can be grounded.

When simplicity and stability are of prime importance, and signal-to-noise ratio is of secondary consideration, the bridge described by Thomas and Huntoon (1949) commends itself. As shown in fig. 12g the specimen circuit is supplied with radiofrequency current through the high resistance $\mathscr{R}_1$ in the same manner as in Rollin's simple circuit. No amplification of the radio-frequency voltage appearing across this circuit is attempted; the voltage is immediately rectified by the crystal valve D_1. Noise and hum modulation in the output is compensated by the output of the other half of the bridge. Here the capacitance potential-dividing combination C_2 and C_3 provides an adjustable radiofrequency voltage which is rectified by the crystal valve D_2 working in opposite sense to D_1. This circuit is compact and gives a measure of the absorption curve uncomplicated by dispersion effects. Moreover, the circuit is relatively free from microphonics since mechanical shocks mainly affect the values of the reactive components of the circuit, and hence to a first order alter only the phase and not the amplitude of the radiofrequency voltage across the specimen circuit. A heavy price is paid for these advantages in the form of a loss by a factor of ten in signal-to-noise voltage ratio which can only be tolerated when χ'' is relatively large.

For his measurements in weak magnetic fields Brown (1950) was able to use the simple Wheatstone bridge arrangement shown in fig. 12h. The condensers C_1 control the amplitude balance, while the condensers C_2 control the phase balance. These experiments were carried out at the relatively low frequencies of 25 and 50 Kc./s., corresponding, for protons, to steady magnetic fields of 6 and 12 gauss respectively.

3.4. Marginal oscillator method

The bridge method, using the circuits of figs. 12a, b, c, d, is a very versatile arrangement for nuclear magnetic resonance measurements on materials of all kinds. The bridge must, however, be kept continually in adjustment, and this makes it inconvenient when searching slowly for unknown lines. In such cases the marginal oscillator method is advantageous.

As with other methods, the specimen is contained in a cylindrical coil placed with its axis perpendicular to the direction of a steady magnetic field. This coil and a condenser form a parallel tuned combination in the grid circuit of a radiofrequency oscillator. Now an oscillator is an amplifier with sufficient positive feedback (regeneration) to supply its own input, and thus maintain continuous oscillation. Formally, a regenerative amplifier may in fact be regarded as placing a negative resistance in parallel with the tuned circuit, so that if this negative shunt resistance more than compensates the positive shunt resistance of the circuit, oscillations are sustained. The amplitude of the oscillations builds up until the average value of the negative resistance over a cycle, as determined by the curvature of the valve characteristic, just equals the positive shunt resistance. Radiofrequency current thus flows in the specimen coil, and when the nuclear magnetic resonance condition is reached an absorption of radiofrequency energy occurs, thus causing a decrease in the positive shunt resistance (as we saw in §3.1). The oscillation amplitude therefore falls every time the modulated 'steady' magnetic field passes through the resonance line. In operation the magnitude of the negative shunt resistance is adjusted to be just small enough to allow oscillations to be sustained. Since the curvature of the valve characteristic is small for such a small voltage swing, the level of oscillation is very sensitive to the small changes in shunt resistance which result from passage through the nuclear resonance condition.

Thus, one reason for restricting the oscillation amplitude to a small level is that the sensitivity is then greatest. A second reason is that at higher levels the valve characteristics are more non-linear, so that noise components originating in a wider band of frequencies are mixed with the nuclear absorption signal, thus increasing the noise factor. The overriding reason, however, is to prevent saturation of the specimen. It is not possible, consistent with stable operation, to reduce the radiofrequency amplitude across the specimen coil much below 0·1 V. While such a level does not generally produce saturation when the specimen is in liquid form, many solids would be saturated. In fact, it is often necessary with solids having a long spin-lattice relaxation time, to work with a radiofrequency level of a few millivolts if saturation is to be avoided; the marginal oscillator method is therefore not suited to the study of such materials.

The audio-modulated radiofrequency voltage appearing across the specimen coil may either be rectified within the oscillator itself by grid rectification, or else may be amplified further and subsequently rectified. The former choice makes the circuit that of an oscillating detector. The grid rectification assists, moreover, in keeping the amplitude of oscillation at a low level; as the amplitude builds up, grid rectification pushes the operating point back to more negative values of grid voltage where the mutual conductance of the valve, and therefore also its gain, is lower. An example of this type of circuit, that of Hopkins (1949), is shown in fig. 13. Almost all standard oscillator circuits can be made to work, but it is

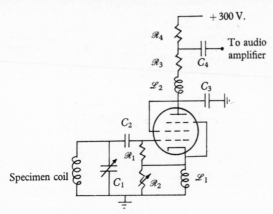

Fig. 13. Hopkins's autodyne circuit, which has a frequency range 8–80 Mc./s. $C_1 = 50$ pF. maximum, $C_2 = 54$ pF., $C_3 = 0.01$ μF., $C_4 = 0.03$ μF., $\mathscr{R}_1 = 510$ K., $\mathscr{R}_2 = 5$ K. maximum, $\mathscr{R}_3 = 22$ K., $\mathscr{R}_4 = 220$ K. R.f. chokes $\mathscr{L}_1$, $\mathscr{L}_2 = 2.5$ mH. Valve is 6 AK 5. $\mathscr{R}_2$ controls feedback.

preferable to avoid circuits such as the Hartley which require three connexions to be made to the specimen coil. Only audio amplification is needed before displaying the signal on an oscillograph or feeding it into a phase-sensitive amplifier. The apparatus is thus very simple and requires no complicated adjustments; this type of circuit is therefore especially useful for lecture demonstrations and for field calibration (see §4.9).

The other choice, namely, of amplifying at radiofrequency before rectification, leads to an improved signal-to-noise ratio, the noise factor being about the same as for the bridge method. Circuits of this kind have been developed by Pound (1947a) and Pound and Knight (1950). Later improvements are reported by Watkins and Pound (1951), and details of the latest arrangement are given by Pound (1952). Their carefully designed circuit is shown schematically in fig. 14. The specimen coil $\mathscr{L}$ and the condenser C form the tuned circuit of a cathode coupled oscillator V1. The radiofrequency voltage developed across this circuit is amplified by V2 (there are actually two untuned radiofrequency stages in the latest circuit), and rectified by V3. The demodulated audio signal is then further amplified and fed to an oscillograph or a phase-sensitive amplifier. It will be seen that the grid voltage of V1a is determined

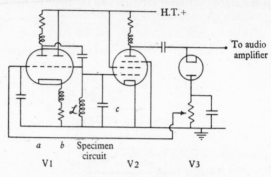

Fig. 14. A schematic diagram of Pound's marginal oscillator circuit for nuclear magnetic resonance measurements.

by the mean detector current, and is so arranged that if the oscillator level should attempt to rise, it is held back by the larger negative grid voltage which results on V1a. The radiofrequency current in the specimen coil may thus be kept close to any desired level above 0·1 V. over long periods of time.

When the object is to discover the resonance line for a nuclear species whose gyromagnetic ratio is at best only roughly known, it is necessary to search for the line slowly, varying the magnetic field at fixed radiofrequency or varying the frequency at fixed field. The instrument then becomes a spectrograph. If the range of search is wide, variation of frequency is usually more convenient, and is accomplished in the Pound circuit by slowly rotating the spindle of condenser C with the aid of a clock motor and a train of reduction gears. With circuits such as that of Hopkins (fig. 13) any appreciable frequency change requires a readjustment to maintain the condition of marginal oscillation. The level of oscillation is, however, automatically maintained in the Pound circuit; by using different specimen coils a frequency range from 1 to 40 Mc./s. is covered.

In searching for a weak line, the resonance must not be swept through in a time less than several times the bandwidth-determining time constant of the phase-sensitive amplifier. Thus for a typical resonance line width of 1 Kc./s. and a time constant of 10 sec., the sweep rate should be about 1 Kc./s. per min. A factor of two in frequency from 6 to 12 Mc./s. would thus take 100 hr. to cover, and it becomes desirable to use 24 hr. per day operation and an automatically recorded output.

Low-frequency noise modulation of the oscillator, caused by flicker effect, is reduced by using a field modulation frequency of several hundred cycles per second instead of the more usual frequency of about 25 c./s. This, however, accentuates another difficulty. Although the specimen coil is set with its axis nominally perpendicular to the 'steady' magnetic field, a small e.m.f. is nevertheless induced in the coil at the modulation frequency. This e.m.f. mixes in the oscillator and produces a spurious amplitude modulation. The solution to this difficulty is to produce a compensating e.m.f. by supplying to an additional coil, wound outside the specimen coil, a current of adjustable amplitude and phase at the modulation frequency.

Other marginal oscillator circuits have been described by Roberts (1947a), Gabillard and Soutif (1950), Knoebel and Hahn (1951) (who use a transitron circuit), Poulis (1951), Kakiuchi, Shono, Komatsu and Kigoshi (1952), Volkoff, Petch and Smellie (1952), Pekárek and Urbanec (1952), Gindsberg and Beers (1953) and Gutowsky, Meyer and McClure (1953). A study of the behaviour of such circuits has been made by Grivet, Soutif and Gabillard (1949), Soutif and Gabillard (1950) and Soutif (1951).

One great advantage which the marginal oscillator methods possess is that they give a pure absorption signal proportional to χ''. The dispersive component of the susceptibility produces a frequency modulation of the oscillator, to which the rectifier does not respond.

3.5. Super-regenerative methods

When the value of $\gamma^2 T_1 T_2$ for a specimen is small, a large radio-frequency field H_1 can be applied to it without producing saturation. In such a case an oscillator using a super-regenerative circuit is sometimes of value, since the circuit is simple and has the property of a good noise factor coupled with the development of a large radiofrequency signal.

A super-regenerative oscillator† is characterized by the repeated build-up and decay of its oscillations. The circuit is made alternately oscillating and non-oscillating by the application of a periodic voltage to one of the electrodes of the oscillator valve. The source of this periodic voltage is usually a separate oscillator, called the 'quench' oscillator.

The coil containing the nuclei under investigation, together with a variable condenser, forms a parallel tuned combination determining the frequency of the radiofrequency oscillator. As we mentioned in the previous section, the oscillator valve effectively places a negative resistance (which is inversely proportional to its mutual conductance), in parallel with the tuned circuit. During the 'on' periods the mutual conductance is large enough to enable the negative resistance to compensate the positive shunt resistance of the tuned circuit. Oscillations build up exponentially until their amplitude is limited by the curvature of the valve characteristic. During the 'off' periods the mutual conductance is rendered low enough to prevent compensation of the positive shunt resistance by the negative resistance; the oscillations then decay exponentially. The time constant for build-up and decay is $2C\mathscr{R}$, where C is the capacity of the condenser tuning the coil, and $\mathscr{R}$ is the resultant shunt resistance of the tuned circuit, which is negative in the 'on' periods and positive in the 'off' periods.

The oscillations build up exponentially from the level of any voltage of the correct frequency developed across the circuit; such a voltage may be either (a) a signal voltage induced in the circuit, or (b) the incompletely decayed oscillation of the previous 'on' period, or (c), in absence of either of these, thermal noise. The quench frequency and the circuit constants are normally adjusted to ensure decay to a level below that of the noise before the next 'on' period, so that (b) is normally absent.

During and just after each 'on' period, an oscillatory radiofrequency current passes through the specimen coil, thus subjecting the specimen to a radiofrequency field. If the coil is correctly oriented in a steady magnetic field whose value is adjusted to the

† For a full account of super-regenerative oscillators see Whitehead (1950).

resonance condition, two effects may occur. First, the absorption of energy by the nuclei causes a reduction of the positive shunt resistance of the tuned circuit, thus increasing the time constant for build-up and reducing the time constant for decay, both producing a reduction in the integrated pulse energy. Secondly, if the period of the quench oscillator is not long compared with the spin-spin interaction time T_2, the coherent precession of the nuclei brought about in the last 'on' period will not have been entirely destroyed at the commencement of the next 'on' period. The nuclear induction e.m.f. generated in the specimen coil by the precessing resultant nuclear moment then acts as a signal voltage from which oscillations build up at the next 'on' period. Since the build-up starts from a higher level than the noise, the integrated pulse energy is increased by this effect. The relative importance of these two effects is determined by the conditions of operation and the nature of the specimen. In general, however, the resonance condition causes a change of the integrated pulse energy, which may be detected either by grid rectification in the oscillator itself, or by reception in a conventional receiver tuned to the oscillator frequency.

The super-regenerative method of detecting nuclear magnetic resonance was first introduced by Roberts (1947a), and has been considerably exploited by Williams (1951) and his colleagues for the measurement of nuclear magnetic moments. Circuits have also been described by Grivet and Soutif (1949), Soutif (1951), Kojima, Ogawa and Torizuka (1951), Suryan (1953), and Gutowsky, Meyer and McClure (1953); the French workers concluded, as did Roberts, that the second of the two effects just discussed is the more important. This does in fact seem likely to be the case since the quench frequency is usually about 10 Kc./s., giving a quench period of about 10^{-4} sec., while for liquid specimens, which have generally been used, T_2 is usually larger than this.

The frequency spectrum of the quenched radiofrequency oscillations consists of a central frequency flanked on either side by a series of sidebands, whose deviations from the central frequency are equal to the quench frequency and its harmonics. In addition to the central frequency, the sideband frequencies may also each satisfy the nuclear magnetic resonance condition. Thus as the steady field is traversed through the region of resonance, a complex

E

pattern of some six equally spaced peaks are observable. The centre of the pattern may be identified by increasing the quench frequency, thus causing the pattern to expand about its centre. This complicated response renders the super-regenerative method unsuitable for experiments other than for the determination of nuclear magnetic moments. Here the complex response becomes the hallmark of the genuine signal when searching for an unknown line; moreover, if the central frequency component should saturate the line, the weaker sideband components may not. The signal-to-noise ratio obtained with this method appears to be quite good, though no direct comparison with other methods has been made.

3.6. Nuclear induction

The experimental arrangements described in the previous sections of this chapter may be termed single-coil methods. In the parlance associated with nuclear induction experiments, these methods have employed a combined transmitting and receiving coil. The received signal frequently produces only a very small amplitude modulation of the transmitted signal; the bridge and oscillator methods were therefore devised to enhance this modulation. The double-coil nuclear induction method devised by Bloch, Hansen and Packard (1946 a, b) and used in their independent discovery in ordinary matter of the nuclear magnetic resonance phenomenon, avoids this difficulty of the single-coil methods.

As we mentioned briefly in §2.8 the nuclear induction signal is received in a second coil placed with its axis perpendicular both to the steady field and to the axis of the transmitter coil. Thus with a perfect geometric arrangement the only signal in the receiver coil is that induced by the precession of the resultant magnetic moment of the nuclei. This signal, which is usually very small, may be amplified considerably before rectification, thus enabling a good noise factor to be obtained. Thus the double-coil specimen circuit achieves the same result by geometric means as do the bridge arrangements by electrical means. It might appear at first sight that the double-coil arrangement achieves this same result without the complication of amplitude and phase controls. This is not the case, however, and, as we shall see presently, there is actually little difference between the methods from this point of view.

A schematic arrangement of a nuclear induction apparatus is shown in fig. 15. The specimen is placed within a receiving coil between the poles of a magnet. The transmitting coil, with its axis

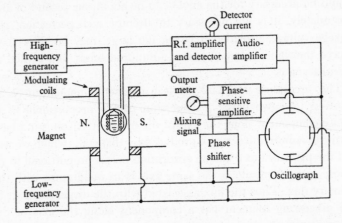

Fig. 15. Schematic diagram of apparatus for the nuclear induction method.

in the third mutually perpendicular direction, is supplied with current from a signal generator. The specimen is arranged to give as good a filling factor as possible for the receiving coil, and the transmitting coil is wound outside it. The receiving coil is tuned by a parallel condenser and forms the input circuit of the first stage of a radiofrequency amplifier. As with the bridge arrangement (fig. 10), the signal may be fed, after rectification and audio amplification, either to an oscillograph or, if weak, to a phase-sensitive amplifier and recording meter. A low audiofrequency source supplies current for modulation of the 'steady' magnetic field, and also provides a synchronous timebase for the oscillograph display and a mixing signal for the phase-sensitive amplifier. In fact the only essential difference between this arrangement and the bridge arrangement lies in the radiofrequency circuit.

We therefore turn now to a closer examination of the double-coil specimen circuit. It will be noticed that if the receiver and transmitter coils are not perfectly orthogonal, a signal is directly induced in the receiver coil from the transmitter. If the angular departure of the receiver coil from perfect setting is ϵ, the directly induced signal, called the leakage, is very roughly equal to a fraction

$\sin \epsilon \cong \epsilon$ of the transmitter voltage. As we saw in §2.1 the nuclear induction voltage may be as low as a fraction 10^{-5} of the transmitter voltage. If the leakage signal were to be made less than this, it would be necessary for the angle ϵ to be about one second of arc. Fortunately, it is not necessary to attempt such perfection; the leakage voltage, provided it is not too large, may be allowed to exceed the signal voltage by, say, a hundred times, and turned to useful account.

The precessing resultant nuclear magnetic moment may be resolved into two parts: a dispersion part u proportional to χ' whose component along the axis of the transmitter coil is in phase with the radiofrequency flux produced by that coil, (which we will call the primary flux), and an absorption part v proportional to χ'' whose component along the same axis is in quadrature with the primary flux. Thus the flux associated with the dispersion part of the precessing moment has a component along the axis of the transmitter coil which is in phase with the primary flux; its component along the orthogonal receiver coil axis is therefore in quadrature with the primary flux. On the other hand, the leakage flux is a part of the primary flux and therefore they have the same phase. The dispersion signal and the leakage signal are thus in quadrature, and the former produces a phase modulation of the latter. On the other hand, the absorption signal is in phase with the leakage signal, and thus produces an amplitude modulation. The detector which follows the radiofrequency amplifier responds only to the amplitude modulation, and thus gives a measure of χ''. Instead of regarding this detector as a mere demodulator of an amplitude-modulated carrier wave, we can in this case regard it as a mixer, which 'heterodynes' the leakage signal with the nuclear induction signal. Since for steady conditions both signals have the same frequency, the mixing is more correctly described as a 'homodyne' process rather than a 'heterodyne' process.

In order to control the leakage flux, Bloch, Hansen and Packard (1946 b) mounted a semicircular sheet of metal which they called a 'paddle', at the end of the transmitter coil, as shown in fig. 16. The paddle causes the flux lines to deviate from an axially symmetrical distribution. By mounting the paddle on a spindle lying along the axis of the transmitter coil, the flux linkage with the receiver coil, whose axis is at right angles in the plane of the diagram, may be varied continuously by rotating the paddle. The leakage flux may thus be reduced to any desired level.

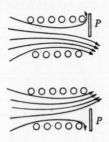

Fig. 16. The steering effect of the paddle P on the lines of radiofrequency flux generated by the nuclear induction transmitter coil, which is shown in section.

We have assumed so far that the leakage flux is merely a part of the primary flux. Any eddy currents produced by the primary flux will, however, generate a leakage flux in quadrature with the primary flux; thus the paddle itself gives rise to flux in quadrature. The paddle alone is therefore insufficient for complete control of the nature of the rectified signal. An additional control was therefore incorporated in two later arrangements (Packard, 1948; Levinthal, 1950). A simple resistance-capacity network feeds a radiofrequency voltage from the transmitter coil to the receiver coil in order to compensate that produced by the quadrature leakage flux. A variable condenser in this network together with the paddle give almost orthogonal adjustment of the signal. If the quadrature component of the leakage flux is completely compensated (by the condenser), and the in-phase component is left incompletely compensated (by the paddle), the absorption signal proportional to χ'' is obtained (v-mode). On the other hand, if the in-phase component is completely compensated, and the quadrature component is incompletely compensated, the dispersion signal proportional to χ' (u-mode) is obtained. The condition of balance is monitored by means of the detector current meter shown in fig. 15, in just the same way as with the bridge circuits.†

† A completely balanced arrangement is described by Baker (1954). A double-coil arrangement, in which balance is achieved solely by electrical adjustments, is described by Gvozdover and Ievskaya (1953).

Proctor (1950) and Weaver (1953) developed the apparatus into a spectrograph to search for lines whose positions are at best only roughly known (cf. Pound's spectrograph, §3.4). Weaver's specimen circuit, shown in fig. 17, controls the in-phase leakage by tilt-

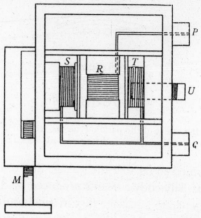

Fig. 17. Simplified diagram of the nuclear radiofrequency mounting used by Weaver (1953). R is the receiver coil within which the specimen is housed. S and T are the two halves of the split transmitter coil; T is fixed, while S may be adjusted by means of the micrometer M. The components are rigidly mounted; internal grounded screens separate off the receiver and transmitter leads to the plugs P and Q, and also the coils from each other, thus minimizing unwanted coupling. U is the u-mode control.

ing the electrical axis of the transmitter coil relative to that of the receiving coil with a micrometer screw,[†] while the quadrature leakage is compensated by the simple inductance loop shown in fig. 18. A high degree of mechanical rigidity is required of the mounting of the two coils in all nuclear induction arrangements, since the leakage flux is very dependent on any relative motion of the coils.

One important piece of information which the nuclear induction methods yield, but which the nuclear magnetic resonance methods previously described do not yield, is the sign of the nuclear magnetic moment. When the leakage signal is adjusted for display of the absorption component χ'' of the susceptibility (v-mode), the absorption signal appears as an amplitude modulation of the leak-

[†] The effect of small deviations of the two coils from orthogonality has been studied by Giulotto, Gigli and Sillano (1947) and Giulotto and Gigli (1947).

age signal, the latter acting as a carrier. Whether the modulation increases or decreases the amplitude of the carrier depends upon whether the absorption component of the nuclear induction signal is in or out of phase with the leakage signal; this phase relationship

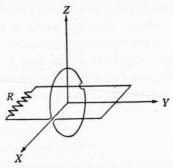

Fig. 18. Weaver's u-mode control, which consists of an inductance loop containing sufficient resistance to make its impedance predominantly resistive. The Y axis is always parallel to the transmitter coil axis; the circular loop therefore picks up a signal which is independent of the orientation of the system about the Y axis. The signal coupled into the receiving coil by the rectangular loop does, however, depend on this orientation, which is adjustable.

thus determines the sign of the absorption signal after rectification. Now the phase of the nuclear induction signal depends upon the sense of precession of the resultant magnetic moment, and thus from equation (2.2) upon the sign of the nuclear magnetic moment. Thus by comparing under identical leakage conditions, the resonance of one nucleus with that for a nucleus whose sign is known, the sign of the first nucleus is found. In practice the requirement of identical leakage conditions implies that the radiofrequency conditions shall be kept constant, and the steady magnetic field changed in order to move from one resonance to the other.

Apart from this advantage of being able to find the sign of the nuclear magnetic moment, which is not important in all types of work, there is little to choose between the single- and double-coil methods of observing nuclear magnetic resonance, and in practice the choice is frequently determined by local experience and tradition. For work at low temperatures one frequently desires simplicity within the cryostat even at the expense of complication outside it. The single-coil methods are preferable in such cases, since only one coil and its two leads are required inside the cryostat, while the nuclear induction methods require two coils, leakage controls and at least three electrical leads within the cryostat.

3.7. Pulse methods

With the exception of the super-regenerative method (§3.5) the experimental arrangements described so far in this chapter have been characterized by the continuous application to the nuclei of a radiofrequency magnetic field at or near their Larmor frequency. Torrey (1949) and Hahn (1950 b) have described experiments in which the radiofrequency field is not applied continuously, but in the form of pulses. Transient nuclear induction signals are obtained both during a pulse and after it. As we shall see in a later section (§5.5) an examination of this transient behaviour enables experimental values of the two relaxation times T_1 and T_2 to be derived.

The apparatus used by both workers is shown schematically in fig. 19, and is a straightforward modification for pulse working of

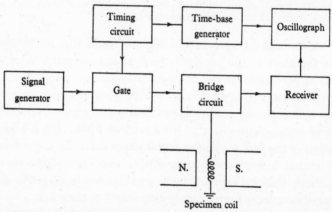

Fig. 19. A schematic diagram of the apparatus for pulse methods.

the bridge methods described earlier. It will be noticed, however, that it is not necessary to modulate the 'steady' magnetic field. The radiofrequency pulses may be obtained either by direct pulse modulation of the radiofrequency oscillator, or as indicated in fig. 19 by gating the output of a continuously oscillating generator. The latter method is preferred, since the frequency is then more stable. Moreover, if it is desired to examine the phase of the transient signal after a pulse has been completed, a source of coherent oscillations is still available for comparison purposes. It is of course necessary to take precautions to prevent any leakage of radiofrequency power through the gate during the 'off' periods.

Torrey (1949), who was mainly interested in the transient response while the pulse was being applied, used the bridge circuit shown in fig. 12 c. Hahn (1950 b), on the other hand, was more interested in the response during the 'off' period after a pulse had ended. In the 'off' period the nuclear induction signal alone is present, so that the bridge circuit is not essential, and the straightforward arrangement of §3.1 may be used. In fact, for these experiments the single coil plays in turn the role of both coils in the double-coil nuclear induction apparatus (§3.6); in the 'on' period the coil acts as transmitter, and in the 'off' period it acts as receiver. Nevertheless, when intense radiofrequency pulses are used, a bridge is desirable to prevent undue overloading of the receiver in the 'on' periods.

Although the pulse methods have mainly been devised for the study of transient effects, Torrey (1949) has pointed out (see also §3.9.1) that their sensitivity is frequently comparable with the sensitivity of the best steady-state methods and that they could therefore be used when searching for unknown resonances.

3.8. Special methods

3.8.1. Optical methods.

In this section we discuss a method of quite a different character from those of the previous sections. We will consider the case of the vapour of the mercury isotope ^{199}Hg, whose nuclear spin number is $\frac{1}{2}$, since this material has been the subject of experimental and theoretical study. The ground state of the mercury atom is $^1S_0$ and for this isotope is split in a magnetic field into two nuclear magnetic sub-levels designated $m = \pm\frac{1}{2}$. The atom may be excited to a $^3P_1$ state by absorption of ultra-violet radiation of wavelength 2537 Å. The quantum number F describing the resultant atomic and nuclear angular momentum can be either $\frac{1}{2}$ or $\frac{3}{2}$ in the excited state. For $F = \frac{1}{2}$ there are two sub-levels $m = \pm\frac{1}{2}$, while for $F = \frac{3}{2}$ there are four sub-levels, $m = \pm\frac{1}{2}$, $\pm\frac{3}{2}$. The level scheme is indicated in fig. 20, where the sub-levels are laid out along the electronic energy levels. In this representation π transitions involving linearly polarized radiation are represented by vertical lines ($\Delta m = 0$), while σ transitions involving circularly polarized radiation are represented by oblique lines ($\Delta m = \pm 1$), the sense of slope being determined by the sense of circular polarization.

Let us suppose the mercury vapour to be placed in a magnetic field and irradiated along the field direction with radiation of wavelength 2537 Å circularly polarized in the sense necessary to produce the transitions shown by the arrows in fig. 20. Atoms in

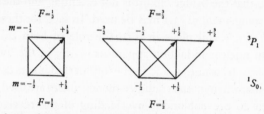

Fig. 20. Scheme showing permitted transitions between the $^1S_0$ ground state and the $^3P_1$ excited state of ^{199}Hg. The arrows mark three particular transitions discussed in the text.

ground-state sub-level $m = +\frac{1}{2}$ can only be raised to the excited sub-level $m = +\frac{3}{2}$, and can only return to ground sub-level $m = +\frac{1}{2}$, emitting σ circularly polarized radiation. Atoms in the ground-state sub-level $m = -\frac{1}{2}$ can only be raised to the excited sub-level $m = +\frac{1}{2}$, but they can return with comparable probability to either the ground-state sub-level $m = -\frac{1}{2}$ by emitting σ radiation or to the ground-state sub-level $m = +\frac{1}{2}$ by emitting π radiation. The exact nature of the polarization of the emitted components depends on the direction of observation relative to the magnetic field, as in the Zeeman effect. It is clear that one result of this optical excitation is to accumulate atoms in the ground-state sub-level $m = +\frac{1}{2}$ by a process of optical 'pumping'. This accumulation is opposed by spin-lattice relaxation processes, for example, by collision, tending to restore the relative population of the sub-levels to that given by the Boltzmann factor. If the intensity of illumination is high enough, and the spin-lattice relaxation time of the vapour is long enough, a considerable disparity of population may nevertheless be achieved, corresponding formally to a very low spin temperature, with a considerable degree of nuclear orientation.

Suppose now that the vapour is, in addition, subjected to a radio-frequency magnetic field of the correct frequency to produce nuclear magnetic resonance between the ground-state sub-levels. The relative population of the sub-levels is now modified, and tends towards an equalization of population as saturation is approached. The population of the ground-state sub-level $m = -\frac{1}{2}$ is thus increased and with it is increased the intensity of the emitted π radiation. By detecting the change of intensity of the emitted π radiation it should therefore be possible to detect (a) the nuclear magnetic resonance condition, and (b) the degree of nuclear orientation produced. This opens up the possibility of detecting nuclear magnetic resonance signals in gases and vapours at low pressure; the methods described in the previous sections usually require a gas pressure of several atmospheres if a detectable signal is to be obtained.

This optical method was suggested by Kastler (1950). Experiments have been carried out on the vapour of mercury [199]Hg at a pressure of the order of 10^{-4} mm. of mercury in a field of 500 gauss by Bitter and Brossel (1952) and by Bitter, Lacey and Richter (1953), though so far without success. They considered that the intensity of optical illumination was high enough and the spin-lattice relaxation time long enough, to ensure a measurable excess of population of the ground sub-level $m = +\frac{1}{2}$. They concluded that the emitted radiation did not escape from the vapour without re-absorption. The exciting radiation may thus be considerably de-polarized, since it does not now consist solely of the incident σ radiation, but also includes re-emitted π radiation. In order to avoid this 'imprisonment' which spoils the optical pumping process, work with a mercury vapour pressure at least as low as 10^{-6} mm. of mercury seems to be required.

It may be mentioned that Kastler's method has been successfully applied to the rather different case of orientation of sodium atoms in an atomic beam (Brossel, Kastler and Winter, 1952; Hawkins and Dicke, 1953), and that the magnetic energy sub-level separations have been explored by application of a radiofrequency field in the manner just described (Brossel, Cagnac and Kastler, 1954). In this case, however, the separation of the magnetic energy sub-levels is not exclusively nuclear.

3.8.2. Aligned radioactive nuclei. It is possible to obtain sufficient alignment of such radioactive nuclei as ^{58}Co and ^{60}Co in certain paramagnetic salts to study the anisotropy of its γ-radiation (Daniels *et al.* 1951; Gorter *et al.* 1951; Daniels *et al.* 1952; Bleaney *et al.* 1954). Such alignment implies that the nuclei predominantly occupy the magnetic sub-levels of greatest $|m|$. Bloembergen and Temmer (1953) point out that if the aligned nuclear spin system were subsequently saturated by electromagnetic radiation at its nuclear magnetic resonance frequency, the nuclear alignment would disappear, and with it the radioactive anisotropy. This disappearance of anisotropy would furnish a means of detecting with precision the resonance condition for suitable radioactive nuclei, and hence would allow the nuclear g-factor to be measured. The advantage of this proposed method is that only a minute quantity of the radioactive isotope would be required, sufficient to give adequate radiation for the measurement of the anisotropy by standard counter technique. If, as in all the experiments mentioned above, alignment is achieved by the method of magnetic hyperfine structure alignment, the nuclei are situated in an intense magnetic field provided by the unfilled electronic shell of the ion; the strength of this field determines the nuclear magnetic resonance frequency.

3.9. General considerations

After discussing in some detail all the main methods for detecting nuclear magnetic resonance, we will now consider two matters which are common to all the methods (except the special methods of §3.8, which will not be discussed further).

3.9.1. Signal-to-noise ratio. If a nuclear magnetic resonance signal is to be detected at all, it must be discriminated from the background of random electrical fluctuations usually called 'noise'; in fact the signal power available at the output terminals of the detection apparatus must be at least of the same order as the noise power. If, further, accurate measurements are to be made of the shape of the resonance absorption line, the ratio of signal power to noise power must be much greater than unity. We will now proceed to evaluate this signal-to-noise power ratio for steady-state detection techniques; this covers the methods of §§3.1, 3.2, 3.3, 3.4 and 3.6.

In §2.8 it was shown that the amplitude $\mathscr{E}_0$ of the nuclear induction e.m.f. in the receiving coil, when adjusted for reception of the absorption or v-mode, is (see equation (2.63))

$$\mathscr{E}_0 = 8\pi\omega\zeta\Lambda\chi''H_1. \qquad (3.15)$$

As before, $\omega/2\pi$ is the radiofrequency, ζ is the filling factor, Λ is the area-turns of the coil, χ'' is the quadrature component of the nuclear magnetic susceptibility, and $2H_1$ is the amplitude of the radiofrequency magnetic field. If the dispersion or u-mode is received, then χ'' must be replaced by χ'. The available power from this source of e.m.f. is therefore

$$\frac{\mathscr{E}_0^2}{8\mathscr{R}_0}, \qquad (3.16)$$

where $\mathscr{R}_0$, as before, is the series resistance of the coil. Although we have spoken in terms of a nuclear induction e.m.f., this result is true of all steady-state systems of detection, since the one-coil and two-coil techniques have been shown to be equivalent.

The expression (3.16) for the available power is only correct if the nuclear induction e.m.f. has a constant amplitude $\mathscr{E}_0$. In practice, however, the modulation technique is used causing a periodic variation of amplitude. Let us suppose that the modulation causes the amplitude of the e.m.f. to vary sinusoidally between zero (away from resonance) and $\mathscr{E}_0$, with a modulation frequency $\omega_m/2\pi$. The e.m.f. $\mathscr{E}$ is then given by

$$\begin{aligned}\mathscr{E} &= \tfrac{1}{2}\mathscr{E}_0(1 + \cos\omega_m t)\cos\omega t \\ &= \tfrac{1}{2}\mathscr{E}_0\cos\omega t + \tfrac{1}{4}\mathscr{E}_0\cos(\omega+\omega_m)t + \tfrac{1}{4}\mathscr{E}_0\cos(\omega-\omega_m)t. \quad (3.17)\end{aligned}$$

The available power in the two sidebands of angular frequency $(\omega \pm \omega_m)$ is therefore

$$\frac{\mathscr{E}_0^2}{64\mathscr{R}_0}, \qquad (3.18)$$

only one-eighth of that given by (3.16). The nuclear induction signal is normally observed in presence of a large amplitude carrier wave on which it forms an amplitude modulation; however, this does not affect the result given in (3.18). The eightfold loss of available power may be pictured as follows. In the first place the modulation is shutting the power off periodically, and for sinusoidal modulation one can easily show that this causes a reduction of the mean power by a factor $\frac{3}{8}$. Secondly, as can be seen from (3.17) only one-third of this remaining power is contained in the sidebands, the remainder causing a modification of the power of the carrier wave.

After linear amplification to a level well above that of the circuit noise, but before rectification has taken place, the ratio of the signal power Π_s to the noise power Π_n is therefore

$$\frac{\Pi_s}{\Pi_n} = \frac{\mathscr{E}_0^2/64\mathscr{R}_0}{kT\mathscr{B}_1\mathscr{F}}, \tag{3.19}$$

where k is Boltzmann's constant, $\mathscr{B}_1$ is the energy bandwidth of the amplifier, and $\mathscr{F}$ is the noise factor of the amplifier.†

After reference to the expression (3.15) for $\mathscr{E}_0$, one sees that the numerator of (3.19) contains the factor $\Lambda^2/\mathscr{R}_0$, which it is more convenient to express in terms of other constants of the specimen coil. To do this we note that if a radiofrequency field of amplitude $2H_1$ were produced by an oscillating current of amplitude $\mathscr{I}_0$ in the receiving coil, the magnetic energy stored in the coil could be written either as $\frac{1}{2}\mathscr{L}\mathscr{I}_0^2$, where $\mathscr{L}$ is the coil inductance, or as $V_c(2H_1)^2/8\pi$, where V_c is the effective volume in which the energy is stored. For a simple solenoid V_c is very nearly equal to the volume of the coil, since the field outside is weak, and the energy is proportional to the square of the field. Equating these two expressions for magnetic energy, and using (2.66) to relate $\mathscr{I}_0$ and H_1, we find

$$\frac{\Lambda^2}{\mathscr{L}} = \frac{V_c}{4\pi}. \tag{3.20}$$

The inductance $\mathscr{L}$ may be expressed as $\mathscr{Q}\mathscr{R}_0/\omega$, where $\mathscr{Q}$ is the quality factor of the coil; making this substitution for $\mathscr{L}$ in (3.20) we have

$$\frac{\Lambda^2}{\mathscr{R}_0} = \frac{\mathscr{Q}V_c}{4\pi\omega}. \tag{3.21}$$

Using this result and equation (3.15), the signal-to-noise power ratio given by (3.19) may now be written as

$$\frac{\Pi_s}{\Pi_n} = \frac{\pi\omega\mathscr{Q}V_c(\zeta\chi''H_1)^2}{4kT\mathscr{B}_1\mathscr{F}}. \tag{3.22}$$

† For a discussion of noise in amplifiers and a definition of noise factor, the reader is referred to Moxon (1949).

We cannot increase this ratio to any desired value by increasing H_1, since χ'', the only factor directly related to the magnetic properties of the material, is a function of H_1, and decreases with increase of H_1 on account of saturation. If our objective is to obtain the largest possible signal-to-noise power ratio, then H_1 must be so adjusted as to give a maximum value to the quadrature component $(2\chi''H_1)$ of the rotating magnetic moment. If we assume for the purpose of calculation that χ'' is given by the Bloch formula (2.54), we find that $(2\chi''H_1)$ is a maximum when ω is equal to ω_0, the Larmor angular frequency, and when H_1 satisfies the condition

$$\gamma^2 H_1^2 T_1 T_2 = 1. \tag{3.23}$$

The maximum value of $(2\chi''H_1)$ is then

$$(2\chi''H_1)_{\max} = \frac{\chi_0 \omega_0}{2\gamma} \sqrt{\frac{T_2}{T_1}} = \tfrac{1}{2}\chi_0 H_0 \sqrt{\frac{T_2}{T_1}}. \tag{3.24}$$

It may be noticed that at exact resonance $(\omega = \omega_0)$, the Bloch expression (2.54) for χ'' and the expression (2.40), taken with (2.16) and (2.25), become identical; expression (2.40) thus leads to the same conclusions if, as in Bloch's theory, one assumes T_2 to be independent of H_1.

Substituting the maximum value of $(2\chi''H_1)$ from (3.24) in (3.22), and putting $\omega = \omega_0$, we get

$$\frac{\Pi_s}{\Pi_n} = \frac{\pi \omega_0^3 \mathscr{Q} V_c \zeta^2 \chi_0^2 T_2}{64\gamma^2 T_1 kT \mathscr{B}_1 \mathscr{F}}. \tag{3.25}$$

The signal-to-noise *voltage* ratio before rectification is just the square root of this power ratio. After rectification the signal-to-noise voltage ratio is of the same order as before detection, and is related to it by a factor $\mathscr{Y}$, which is of the order of unity and depends upon a number of factors, including the detector law. We may broaden the meaning of this factor $\mathscr{Y}$ to take account at the same time of any inexactitude in our earlier supposition that the process of modulation produces a *sinusoidal* variation of nuclear induction e.m.f. between o and $\mathscr{E}_0$. The actual waveform depends on the modulation amplitude and waveform, and on the shape of the absorption line (cf. Grivet 1951). Since only energy in the first sidebands of angular frequency $(\omega \pm \omega_m)$ is utilized by the phase-sensitive amplifier, $\mathscr{B}_1$ in (3.25) must now be replaced by $2\mathscr{B}_2$, where $\mathscr{B}_2$ is the effective bandwidth of this amplifier.

Taking the square root of (3.25), replacing $\mathscr{B}_1$ with $2\mathscr{B}_2$, substituting for χ_0 from (2.33), and inserting the factor $\mathscr{Y}$, we find the following expression for the signal-to-noise voltage ratio:

$$\frac{\mathscr{V}_s}{\mathscr{V}_n} = \left(\frac{\mathscr{Y}\zeta N\gamma I(I+1)h^2}{48kT}\right)\left(\frac{V_c \mathscr{Q}\nu_0^3 T_2}{kT\mathscr{B}_2\mathscr{F}T_1}\right)^{\frac{1}{2}}. \tag{3.26}$$

This expression, which was derived by Bloembergen, Purcell and Pound (1948), leads to the following conclusions:

(a) *Frequency* ν_0. The frequency occurs explicitly as $\nu_0^{\frac{3}{2}}$, but also enters implicitly into $\mathscr{Q}$ (which usually increases slowly with frequency), $\mathscr{F}$ and V_c (which may slowly decrease with increase of frequency). It may be concluded that it generally pays to work at as high a frequency as possible. This frequency is frequently limited by the magnetic field available, which should therefore be as high as possible.

(b) *Gyromagnetic ratio*. This occurs explicitly as γ, but if the frequency of operation is limited by the available field, then on substituting from (2.3) an additional explicit factor $\gamma^{\frac{3}{2}}$ appears. The signal-to-noise ratio is thus highly favoured by a large γ.

(c) *Number of nuclei present*. The proportionality of signal-to-noise ratio to $\zeta NV_c^{\frac{1}{2}}$ indicates that the specimen should contain as many resonating nuclei as possible.†

(d) *Relaxation times*. The factor $\sqrt{(T_2/T_1)}$ implies that the line width should be small, as for example in a liquid, and that the spin-lattice relaxation time should be short. T_1 may often be shortened by the addition to the specimen of a paramagnetic impurity.

(e) *Detection system*. The factor $(\mathscr{Q}/\mathscr{B}_2\mathscr{F})^{\frac{1}{2}}$ leads to the obvious statement that the signal-to-noise ratio is improved by use of a high-$\mathscr{Q}$ coil, an amplifier of low noise-factor, and a narrow bandwidth. As was mentioned in §3.2, $\mathscr{B}_2$ is of the order of $1/t_c$, where t_c is the time constant of the output circuit of the phase-sensitive amplifier; $\mathscr{B}_2$ is therefore only reduced at the expense of a slower speed of recording information.

---

† The proportionality of signal strength to N suggests incidentally that the resonance signal strength could be used to measure the proportion of a given isotope in a sample of the element, and to monitor the progress of isotope extraction.

As an example of the use of (3.26), we find that the signal-to-noise voltage ratio for the proton resonance in 1 cm.^{3} of pure water in a field of 5000 gauss using a bandwidth $\mathscr{B}_2$ of 1 c./s. is about 1000. The width of the absorption line, and therefore also the value of T_2, is here assumed to be entirely due to a field inhomogeneity of 0·1 gauss. If the signal is displayed on an oscillograph, an audio-amplifier bandwidth $\mathscr{B}_2$ of say 5 kc./s. may be required, thus reducing the signal-to-noise voltage ratio to about 15. It is thus seen that the ratio may be very weak for nuclei of low gyromagnetic ratio, or for nuclei present in small concentration as in a gas, or for many solids, which have long T_1 and short T_2.

It may be asked whether the signal-to-noise ratio for the dispersion or u-mode differs from that of the absorption or v-mode treated so far. Using the Bloch formula (2.53) for χ', the maximum value of $(2\chi' H_1)$ is found to be the same as is given by (3.24) for $(2\chi'' H_1)$, and is approached asymptotically for large H_1. There is, however, one small difference, in that the minimum value of $(2\chi' H_1)$ has the same magnitude as the maximum, but with opposite sign, whereas the minimum value of $(2\chi'' H_1)$ is zero. Thus, as H. L. Anderson (1949) has pointed out, an additional factor of 2 should be inserted in the numerator of (3.26) when the dispersion signal is observed.

For his pulse method, Torrey (1949) has shown that the maximum value of $(2\chi'' H_1)$ at the commencement of a radiofrequency pulse is $\chi_0 H_0$, the static magnetic moment, instead of the smaller steady-state value given by (3.24). This results in an improvement in available signal power by a factor $4T_1/T_2$, which may well be as large as 10^4. On the other hand, in order to make use of this available power, the receiver bandwidth must be wide enough to pass the pulses without undue distortion, and is usually several kilocycles per second. The increase in available noise power thus largely offsets the increase in signal power, and the two methods have comparable sensitivity. It is in any case difficult to make a precise comparison of sensitivity of pulse and steady-state methods, since a psycho-physiological perception factor must be applied to the final signal-to-noise voltage ratio, and this factor may differ considerably with the different methods of presenting the final information.

F

3.9.2. Magnet. The intrinsic width of a nuclear magnetic resonance absorption line, expressed in gauss, is usually small in comparison with the value of the steady magnetic field. This is particularly true for specimens in liquid or gaseous form. If this narrowness is to be fully exploited when making a precise determination of the resonance condition, or if it is desired to make a detailed examination of the resonance line shape, it is essential that the steady magnetic field shall be as homogeneous as possible throughout the volume occupied by the specimen. Moreover, the field strength must remain as nearly constant as possible throughout the duration of any series of measurements. The properties of constancy and homogeneity are in fact the two most important requirements of a magnet for work of this kind, and we shall consider each in turn.

The field of an electromagnet supplied with direct current from a motor generator is not usually sufficiently constant for precise work. An adequate improvement is frequently obtained by replacing the generator with a large capacity battery of storage cells. A much closer control of the field of the electromagnet may be obtained at the expense of complication by use of an electronic regulator. These are of two types; either the magnet current is held constant or the field itself is held constant. The two types are not quite equivalent in their results, since temperature changes in the electromagnet produce small geometrical changes and changes of permeability of the iron, thus slightly altering the field even though the current remains constant. Regulators of the second type making use of the nuclear magnetic resonance phenomenon itself are discussed later in §4.10.

Another widely used means of producing a constant field is the permanent magnet. The constancy is comparable with that of the electronically regulated electromagnet, while at the same time it is simpler to use, easier to maintain, and consumes no power. The field strength is somewhat temperature-dependent, though the large heat capacity of the magnet prevents rapid changes. It is found in practice that with auxiliary coils one can vary the field almost linearly with current over a range of at least ± 25 gauss, which is sufficient for most experimental purposes. The main disadvantage of the permanent magnet is its lack of flexibility. The field must usually be accepted as constant, so that one must vary the operating frequency if one is to work with the resonance of different nuclear species. Permanent magnets designed for nuclear magnetic resonance work have been described by Pake (1948), Pound (1950), Gutowsky and Hoffman (1951), Andrew and Rushworth (1952) and Gutowsky, Meyer and McClure (1953).

In order to meet the second requirement of field uniformity, most workers have used accurately-ground plane-parallel circular polecaps made of soft iron which has been carefully treated to ensure homogeneity of its magnetic properties. The diameter of the polecap is usually made at least three times the width of the gap in order to reduce the inhomogeneity of the field which results from the radial fall of field towards the edges of the gap. This edge effect may be very considerably reduced by the use of a ring shim of the type suggested by Rose (1938), who gives the necessary dimensions of the shim for cylindrical polecaps. When a permanent magnet is employed, efficient design usually requires the polecap to have the form of a truncated cone. The dimensions of the necessary ring shim in this case have been calculated by Andrew and Rushworth (1952). For a specimen of 1 cm.^{3} volume, a homogeneity of field of a few parts in 10^5 may be obtained without much difficulty, and this may be further improved by hand-polishing the polefaces (Gutowsky and Hoffman, 1951). Other workers (e.g. Wimett, 1953b) have achieved a high degree of field uniformity by the use of empirically shaped polefaces.

CHAPTER 4

MEASUREMENT OF NUCLEAR PROPERTIES AND GENERAL PHYSICAL APPLICATIONS

4.1. Determination of nuclear magnetic moments

In §2.1 it was shown that the nuclear magnetic resonance condition is (see equation (2.1))

$$h\nu_0 = \frac{\mu}{I} H_0. \tag{4.1}$$

From this equation it is seen that one can determine the magnetic moment μ of a nucleus by measuring its magnetic resonance frequency ν_0 in a known magnetic field H_0, provided of course that one knows the nuclear spin number I, and that one has an adequate value for Planck's constant h.

As will be seen presently, the measurement of μ can be made with considerable accuracy by this method; only the molecular beam resonance method, which requires a more complicated experimental technique, has a comparable accuracy. The measurement of nuclear magnetic moments has in fact proved to be the most important application of the nuclear magnetic resonance phenomenon, and the moments of more than seventy species of nuclei have already been determined by this method. This compilation of accurate nuclear properties provides essential data for those attempting to formulate an adequate theory of nuclear structure.

From equation (4.1) it is seen that the absolute determination of μ requires a knowledge of the four other quantities h, ν_0, H_0 and I, and it is instructive to consider the relative accuracy with which the first three of these are known. A radiofrequency is one of the most accurately measurable physical quantities; indeed, by using experimental specimens in liquid form, and having a very homogeneous magnetic field, the resonance signal is often sharp enough to allow the resonance frequency to be determined with an absolute accuracy of better than one part in 10^6. By contrast, Planck's constant is only known with an accuracy of the order of one part in 10^4, while the measurement of a magnetic field to an accuracy better than one part in 10^3 is a major undertaking. Consequently in order to make the fullest use of the great precision possible in determining the resonance frequency, the large majority of the measurements made have been concerned with the comparison of magnetic moments rather than their absolute determination. Thus if the resonance frequency for two species of nucleus a and b are determined in the same magnetic field, then from (4.1) we see that

$$\left(\frac{\mu}{I}\right)_a \Big/ \left(\frac{\mu}{I}\right)_b = \frac{(\nu_0)_a}{(\nu_0)_b}. \tag{4.2}$$

Thus, assuming that the spin numbers are known, the measured frequency ratio can give the ratio of the magnetic moments with great accuracy.

As we shall see in §§4.5, 4.6 and 4.7, the magnetic moment of the proton has been determined by precision methods in (a) absolute units, (b) nuclear magnetons and (c) Bohr magnetons. If therefore the proton is taken as one of the nuclei a, b in the comparison experiment, the magnetic moment of the other nucleus can be expressed in any of the above units. The proton has in fact often been used as a reference nucleus in this manner, though sometimes it has been found more convenient to use an intermediate reference nucleus, for example the deuteron, ^{7}Li or ^{23}Na, which has itself been compared with the proton.

An appreciable loss of accuracy is usually suffered in converting from the observed frequency ratio to the value of the desired magnetic moment, since the frequency ratio can generally be obtained with greater accuracy than that with which the proton moment is known. It is therefore of interest to note that an accurate value of the frequency ratio alone is of value for a pair of isotopes of the same element, since such a ratio can sometimes be combined with observations of the hyperfine structure of the atomic spectrum of the element to demonstrate the effect of finite nuclear size on the hyperfine splitting (Bitter, 1949 b; Crawford and Schawlow, 1949; Bohr and Weisskopf, 1950; Gutowsky and McGarvey, 1953 a; Brun, Oeser, Staub and Telschow, 1954 a; Sogo and Jeffries, 1954). Furthermore, as will be seen later in this section, the measured frequency ratio for such pairs is immune from several corrections which normally have to be applied.

It has been assumed so far that the nuclear spin number I is known and can be inserted in (4.1) and (4.2). If it is not already known, it can often be determined by nuclear magnetic resonance methods, as is shown in §4.3. If, however, there is no certain information as to the value of I, then instead of being able to quote the value of the nuclear magnetic moment, one must be content with quoting the value of the nuclear g-factor, since, as was seen in §2.1, gI is the value of the nuclear magnetic moment in nuclear magnetons.

In order to measure the frequency ratio required in equation (4.2), it is first necessary to give careful consideration to the choice of specimens containing the nuclei of interest and the reference nuclei. Specimens in solid form are usually undesirable, since their resonance line is normally rather broad (§2.3), thus reducing the precision with which the centre of the line may be located. Gaseous specimens give sharp resonance lines but are usually inconvenient, since the gas must generally be compressed in order to obtain an adequate signal strength. For the inert elements, however, a gaseous specimen cannot be avoided at ordinary temperatures (e.g. ^{3}He (Anderson, 1949) and ^{129}Xe (Proctor and Yu, 1951)). Most work has therefore been done with specimens in liquid form, since strong narrow signals may then be obtained. Such specimens may either consist of a compound of the desired element in liquid form or of a solution in which such a compound is the solute.

It is frequently necessary to reduce artificially the spin-lattice relaxation time T_1 of the specimen; this allows the radiofrequency field amplitude to be increased without saturation, and thus improves the signal-to-noise ratio for otherwise weak signals. For specimens in the form of aqueous solutions this is readily achieved by dissolving a small amount of paramagnetic salt, since the paramagnetic ions promote the relaxation process (see §5.4). It is, however, important that the relaxation time should not be reduced too far, since, as mentioned in §2.3 (see also §5.3), this causes an undesirable broadening of the resonance line, thus limiting the precision with which the centre may be located.

If the specimen is a solution, the solute must be very soluble and mutually soluble with any added paramagnetic salt. The compound must not contain unpaired electrons, since otherwise the nuclei of interest are subjected to intense internal magnetic fields. If the nucleus has an appreciable electric quadrupole moment it should be situated in a position of high symmetry in the molecule or ion in order to reduce the electric field gradient with which the quadrupole moment interacts, thus avoiding undue broadening of the resonance line by quadrupole spin-lattice relaxation (§8.2).

Having chosen the specimen it is next necessary to locate the resonance signal. If the nuclear magnetic moment is already fairly well known from molecular beam work, this presents little difficulty. On the other hand, if its value is only roughly known from optical hyperfine structure work (§1.1), or is merely estimated on the basis of empirical rules, it is necessary to search slowly and carefully over a wide range of frequency. Automatic spectrographs of the kind described in §§3.4 and 3.6 are particularly convenient for this. Once found it is necessary to compare the resonance frequency with that for the reference nucleus in the same magnetic field. Here a number of procedures are possible. A specimen containing the reference nuclei may be substituted for the original specimen and the new resonance frequency found using the same radiofrequency equipment. Alternatively, the two specimens may be mounted side-by-side and their resonance frequencies measured simultaneously with two parallel sets of equipment, one for each specimen. In the first procedure a correction has to be made for any change of magnetic field in the time interval between consecutive measurements, and in the second for any difference in field at the two specimen sites. A better procedure, where it can be arranged, is to prepare a single specimen containing both the nuclei of interest and also the reference nuclei. Two radiofrequency coils may be wound orthogonally round the mixed specimen, and two parallel sets of equipment can be arranged to record the two resonance signals simultaneously, thus avoiding either correction. Alternatively, the two resonances could be recorded consecutively for the mixed specimen using one set of equipment, but with a much shorter time elapsing between measurements than when specimens have to be interchanged.

A number of corrections have now to be applied to the measured frequency ratio before it can be inserted in (4.2) and used to find the magnetic moment of the nucleus of interest. Although the same magnetic field has been applied to both specimens (or a correction made if the fields were not identical), the field actually experienced by the nuclei differs somewhat from the applied field, and by an amount which in general differs for the two types of nucleus. The necessary corrections have been considered in detail by Dickinson (1951) and will be mentioned briefly in order:

(1) *Correction for bulk diamagnetism of the specimen.* First of all we consider a specimen containing no paramagnetic ions, so that the bulk susceptibility χ_1 is diamagnetic. In order to find the field acting on a particular molecule or ion we use the method well known in dielectric theory (see Debye, 1945). The contributions are considered separately from the portion of the specimen lying within a small macroscopic spherical surface centred upon the molecular ion, and from the portion lying outside. For a liquid, Lorentz (1909) finds that the contribution of the portion within the spherical surface may be taken as zero. The outer portion produces within the spherical cavity a field greater than that in absence of the cavity by a fraction $\frac{4}{3}\pi\chi_1$. However, the field in the specimen is less than the externally applied field by a fraction $\kappa\chi_1$, where κ is the demagnetization factor of the specimen which depends upon its shape. (κ is $\frac{4}{3}\pi$ for a sphere, 2π for a long cylinder transverse to the field, and zero for a long cylinder parallel to the field.) Thus the molecule or ion finds itself in a field

$$[1 + (\tfrac{4}{3}\pi - \kappa)\chi_1]H_0 \qquad (4.3)$$

rather than H_0. If the specimen is spherical the correction factor is seen to disappear. If a mixed specimen is used, the factor is the same for both types of nucleus, and does not affect the frequency ratio. If, however, two specimens are used, of different κ and χ_1, a correction is necessary. Since $|\chi_1| \sim 10^{-6}$ for most diamagnetic liquids, the correction is only of importance in work of the highest precision.

(2) *Correction for bulk paramagnetism.* If the specimen contains paramagnetic ions the bulk susceptibility χ_1 may be of the order 10^{-4} to 10^{-5}, and the correction becomes quite important.† Un-

---

† The static paramagnetic susceptibility χ_0 due to the nuclear magnetic moments themselves, given by equation (2.33), is only of the order 10^{-10}, and may therefore be neglected.

fortunately, Dickinson (1951) has found experimentally that in this case the factor $(\frac{4}{3}\pi - \kappa)$ of (4.3) is in general not correct, and that the actual factor may be several times larger or smaller, and may even be of opposite sign. A suggested explanation of this anomaly has been put forward by Bloembergen and Dickinson (1950),[†] who consider that the contribution to the local field from within the spherical surface is in this case not zero, as we assumed above. Paramagnetic ions behave anisotropically in the presence of electric fields, and although the average electric field at any point in a liquid is zero, the instantaneous value is not. Consequently the average local magnetic field from the instantaneously anisotropic ions within the sphere is not in general zero. If this explanation is correct one would expect the change of local field to depend upon the chemical form of the specimen rather than on the nucleus of interest. In confirmation of this, Dickinson finds that the change of local field is the same for a pair of isotopes of the same element when in the same chemical combination, but is different for the same isotope in different chemical combination. The use of a mixed specimen does not therefore eliminate this correction unless the two nuclei are isotopes of the same element.

Since this effect cannot at present be exactly predicted it is therefore preferable to avoid solutions containing paramagnetic ions. If they cannot be avoided, the ion concentration should be as weak as possible, and its exact value stated when measurements are published, so that correction may be possible in the future.

(3) *Magnetic shielding by the electronic system of the atom.* Having discussed the value of the magnetic field that is actually applied to the ion or molecule, we must now consider the diamagnetic screening effect of the orbital electrons upon the field at the nucleus itself. Lamb (1941) has derived a formula for this correction for a free atom or monatomic ion. For its evaluation a knowledge is required of the electrostatic potential produced at the nucleus by the atomic electrons, and Lamb used the Fermi–Thomas atomic model to produce an explicit formula for the correction for all elements; this correction increases from a fraction of about 3×10^{-5} for hydrogen to about 0·01 for the heaviest elements. This correc-

[†] An explanation along the same lines has since been advanced by Grivet and Ayant (1951) and Ayant (1951 a, b).

tion has been further considered by Dickinson (1950 b) and Proctor and Yu (1951).

Although this correction is thus calculable with some confidence for a free atom or monatomic ion, the situation is less satisfactory for molecules or complex ions. Ramsey (1950a, b, 1951, 1952b) has found that the shielding correction in this case consists of a diamagnetic term not very different from that just discussed for the free atom or ion, together with a second-order paramagnetic term.

The calculation of this second term, which is by no means negligible, cannot usually be carried out since it requires a knowledge of the wave functions for the excited states of the molecule, and these are in general not known. This second-order paramagnetic term differs with the molecule in which the nucleus is found and is a contributory cause, if not the main cause, of the chemical shift about which more will be said in §5.6. For the important case of gaseous molecular hydrogen, Ramsey (1950a, b, 1951) and Hylleraas and Skavlem (1950) have been able to calculate the correction. However, water and mineral oil are more widely used as specimens for the proton resonance than hydrogen gas; the very small differences in shielding correction between these liquid specimens and the gas have been found experimentally from direct comparisons of their resonance frequencies carried out by Thomas (1950) and by Gutowsky and McClure (1951).

Since, apart from the case of hydrogen, the chemical shift cannot at present be calculated, this indicates yet another reason why it is important that full details of the specimens used should be given when the results of measurements are published.

(4) *Oscillating field correction.* As was seen in §2.1 a linearly oscillating field may be regarded as the superimposition of two oppositely rotating radiofrequency fields, only one of which is utilized in magnetic resonance. Bloch and Siegert (1940) have shown that the other component has the effect of shifting the resonance value by factor of the order $(H_1/H_0)^2$, where, as before, $2H_1$ is the amplitude of the linear oscillating radiofrequency field. Since the ratio H_1/H_0 rarely exceeds 10^{-4} in practice, this correction is quite negligible.

A list, with references, is given in Appendix 2 of all the nuclei whose magnetic moments have been measured by the nuclear magnetic resonance method. In view of its theoretical importance, the greatest experimental effort has been devoted to the measurement of the nuclear magnetic moment of the deuteron relative to that of the proton. The use of a mixed specimen eliminates the first correction (for bulk diamagnetism); moreover, for a pair of isotopes of the same element the second correction (for bulk paramagnetism) is also eliminated even if paramagnetic ions should be present in the specimen; for the same reason the third correction (for magnetic shielding) is eliminated to a first order. There is, however, a second-order difference in the shielding correction which for most isotope pairs is negligible, but which is of particular importance in the case of hydrogen. As discussed by Smaller (1951) and Newell (1950), the large difference in the masses of the proton and the deuteron gives rise to a difference in the zero-point vibration amplitude of the hydrogen molecule and consequently to a difference in the magnetic shielding even in molecules which are identical except for the interchange of deuteron and proton. Ramsey (1952a) has therefore pointed out the desirability of measuring both the proton and the deuteron resonance in the molecule HD. In this molecule the modification of the magnetic shielding by the zero-point vibration, which is at present not calculable, is the same for both nuclei. Such a measurement has recently been made (Wimett, 1953 b).

4.2. Determination of the sign of nuclear magnetic moments

As was discussed in §3.6, one special virtue of the nuclear induction technique is that it gives information concerning the sign of the nuclear magnetic moment, whereas the single-coil methods do not. The sign determines the sense of precession of the nuclear magnetic moment in a magnetic field, and hence also the phase of the nuclear induction signal relative to the leakage signal. This in turn determines whether the signal shall be represented on the oscillograph by an upward or a downward displacement of the trace. In order to know which direction of displacement represents a positive sign and which a negative sign, the signal for a nucleus of known sign is displayed under identical leakage conditions. Since the leakage conditions are liable to change with change of radiofrequency, the signal from the reference nucleus is obtained by changing the value of the steady magnetic field, leaving all other conditions the same.

The nuclear magnetic moments whose signs have been determined or confirmed in this way are indicated in Appendix 2. Strictly the method only compares the signs of the two nuclear magnetic moments, finding whether they are like or unlike, and ultimately relies upon a knowledge that the sign of the proton moment is positive in order to give actual signs. That the sign of the proton moment is positive is known from molecular beam work (Kellogg, Rabi, Ramsey and Zacharias, 1939), and has more

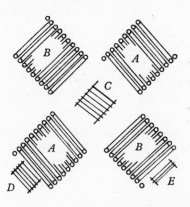

Fig. 21. The nuclear induction arrangement of Staub and Rogers (1950) for determining the sign of the proton magnetic moment. *AA* and *BB* are split transmitter coils supplied with radiofrequency current in quadrature, thus producing a rotating magnetic field at the centre of the system. *C* is the receiving coil containing the specimen. *D* and *E* are compensating coils in series with *C* to provide cancellation of signals directly induced by the transmitter coils. The steady magnetic field is directed normal to the plane of the diagram.

recently been confirmed by a nuclear induction method (Rogers and Staub, 1949; Staub and Rogers, 1950). In the latter experiment a rotating radiofrequency field was generated rather than a linearly oscillating field. This rotating field was obtained by means of crossed transmitter coils (see fig. 21) supplied with radiofrequency current differing by 90° in phase. The sense of rotation of the radiofrequency field was determined from a comparison on an oscillograph of the phases of the current in the crossed transmitter coils.

4.3. Determination of nuclear spin numbers

Before the advent of the nuclear magnetic resonance methods, the spin numbers of the majority of the nuclei to which these methods could be applied had already been determined by other methods, as, for example, from band spectra, from the hyperfine structure of line spectra and from atomic beam work. Nevertheless, some spin numbers have been determined for the first time by the new methods, and many more have been confirmed; these are indicated in Appendix 2. Several methods may be used, and these will be considered in turn.

4.3.1. Signal strength method.
This method has been used more extensively than the others. Suppose we measure the ratio of the maximum amplitude of the resonance signals, for example as displayed on an oscillograph, for two species of nucleus a and b under the same conditions. The same radiofrequency coil is used; the same radiofrequency is used and H_0 is varied in order to move from one resonance to the other; the amplitude $2H_1$ of the radiofrequency field is held constant, and at a low enough level to avoid saturation in both cases (i.e. so that we may put the saturation factor Z equal to unity); the apparatus is maintained in adjustment for observation of the absorption mode proportional to the quadrature component χ'' of the nuclear magnetic susceptibility; and finally, the temperature is held constant. Then using expression (3.15) for the nuclear magnetic resonance signal amplitude, and substituting in it for χ'' from (2.40) and from (2.33) for χ_0, and expressing μ as $\gamma I \hbar$, we find for the ratio of the resonance signal voltages

$$\frac{\mathscr{E}_{0_a}}{\mathscr{E}_{0_b}} = \frac{\zeta_a N_a \gamma_a^2 I_a (I_a + 1) g_a (H_{0_a})}{\zeta_b N_b \gamma_b^2 I_b (I_b + 1) g_b (H_{0_b})}, \tag{4.4}$$

where the subscripts a and b refer to the two species of nucleus, and where the normalized shape function g has been written as a function of H_0 rather than of the ν, since in this experiment the value of the steady field is varied rather than the frequency. The ratio given in (4.4) is actually the ratio of the signal voltages at the terminals of the radiofrequency coil containing the specimen; it is also equal to the ratio of the signal voltages as displayed on the oscillograph if the intervening stages of amplification and detection are linear.

It is seen from (4.4) that the ratio of signal strengths is markedly dependent on the values of the spin number I, and that if the spin number is known for one of the two types of nucleus, then the other spin number can readily be found from this equation. The ratio $(\zeta_a N_a)/(\zeta_b N_b)$ is the ratio of the numbers of nuclei of the two types present, since ζ is the filling factor and N the number of nuclei per cm.[3]; the ratio γ_a/γ_b is just the ratio of the frequencies at which the two resonances may be observed in the same steady field; while the ratio g_a/g_b at the peak value of each resonance, which from (2.16) equals T_{2a}/T_{2b}, is also the inverse ratio of the widths of the resonance signals, provided they have geometrically similar shapes. If the shapes are not approximately similar it is not sufficient just to find the ratio of the maximum signal strengths, but instead it is necessary to find the ratio of the integrated experimental line shapes, i.e. the ratio of the areas under the displayed signals; carrying out this integration, and remembering that the line-shape functions g are normalized, we obtain the following expression in which the shape functions no longer appear:

$$\frac{\int_0^\infty \mathscr{E}_{0_a}\, dH_{0_a}}{\int_0^\infty \mathscr{E}_{0_b}\, dH_{0_b}} = \frac{\zeta_a N_a \gamma_a^2 I_a (I_a + 1)}{\zeta_b N_b \gamma_b^2 I_b (I_b + 1)}. \tag{4.5}$$

In practice it is not necessary to measure the quantities in (4.4) and (4.5) with much precision, since the spin number must be either an integer or half-integer. Thus, for example, if the mass number of the nucleus a whose spin number is to be determined is odd, then I_a must necessarily be one of the series $\frac{1}{2}, \frac{3}{2}, \frac{5}{2}, \ldots$, so that the quantity $I_a(I_a + 1)$, which appears in (4.4) and (4.5), must be one of the series of well-separated quantities 0·75, 3·75, 8·75, ... Similarly, if the mass number is even, I_a must be an integer, and $I_a(I_a + 1)$ must be one of the series 2, 6, 12, ... Thus quite a modest accuracy normally suffices to identify the value of I_a.

If the resonance line shape is at all complicated, it is essential to find the area under the curve and apply (4.5), rather than relying upon the ratio of maximum values in conjunction with (4.4). It is, moreover, important to be sure that the whole of the resonance curve has in fact been integrated. This moral is pointed by a case reported by Sands and Pake (1953) in which for ^{29}Si the product $I_a(I_a + 1)$ appeared to be $2 \cdot 0 \pm 0 \cdot 3$, thus lying well between the first two members $0 \cdot 75$ and $3 \cdot 75$ of the series of permitted values for a nucleus of odd mass number. Watkins and Pound (1953), who had found similar anomalies, suggested that the ^{127}I resonance in solid potassium iodide, with which the ^{29}Si resonance had been compared, was about three times smaller than its proper value, thus bringing the true value of $I_a(I_a + 1)$ down to about $0 \cdot 7$ and leading to a spin number of $\frac{1}{2}$ for ^{29}Si. They suggested that internal strains within the potassium iodide crystal had destroyed the cubic symmetry of the electric field acting upon the electric quadrupole moment of the ^{127}I nucleus, and that in consequence (as is discussed further in §8.3) the resonance was broken up into a central line with two satellites on each side. The satellite lines were broad and weak and so escaped observation, with the result that only the central line, which in this case encloses only about a third of the total integrated area, was taken into account.

Although it is frequently more convenient to find an unknown spin number by a comparison of the resonance signal strength with that of a nucleus of known spin number, a measurement may alternatively be made of the absolute signal strength, and the spin number found using equation (3.15), (2.40) and (2.33). For this purpose Pound (1952) includes a signal strength calibrator in his spectrograph circuit.

4.3.2. Electric quadrupole splitting method. In §2.1 we saw that for a nucleus of spin number I which is placed in a magnetic field, there were available to it $(2I + 1)$ equally spaced energy levels, one level for each of the $(2I + 1)$ permitted values of the magnetic quantum number m. If $I > \frac{1}{2}$, the nucleus has in general an electric quadrupole moment. As is discussed more fully in §8.3, such a quadrupole moment interacts with any static electric field gradient which may exist at the position of the nucleus, producing a displacement of the energy levels. The magnitude of the displacement is a function of m, with the result that the $(2I + 1)$ levels are no longer equally spaced, and the $2I$ intervals between them are now in general all unequal. Thus instead of finding a single resonance line, $2I$ lines are found instead. It is therefore only necessary to count the number of lines in such a resonance spectrum to find I.

In order to apply this method it is necessary to choose a specimen in solid form if the electric field gradient is to be static. The material chosen must contain the nuclei in an asymmetrical environment within the crystal lattice to ensure the presence of an inhomogeneous electric field. All the nuclei of interest should preferably occupy equivalent positions in the lattice, so that all are subjected to the same electric field gradient. Furthermore, since the separations between the $2I$ component lines also depend upon the direction of the steady magnetic field H_0 with respect to the crystal axes, it is necessary to use a single crystal to ensure unambiguous results.

4.3.3. Other methods.

It was mentioned in §2.3 that the resonance line for a solid specimen may be several gauss broad on account of the interaction between the neighbouring nuclear magnetic dipoles. It will be shown in §6.2.3 that the contribution of a given nuclear neighbour to the mean square width of the resonance line is proportional to $I(I+1)$, where I is the spin number of the neighbour. It is therefore possible with a suitable specimen to find I from the observed mean square width. Gutowsky, McClure and Hoffman (1951) have applied the method to ^{9}Be.

Another method makes use of the fine structure of the resonance line found in certain liquid specimens (see §6.7), which arises from magnetic interaction between nuclei in the same molecule via the molecular electrons. The number of lines in the fine-structure pattern is dependent upon the number of nuclear neighbours in the same molecule and their spin number, and in suitable cases can lead to the determination of the spin number of a particular nuclear neighbour. The method has been applied to ^{29}Si by Ogg and Ray (1954) and Williams, McCall and Gutowsky (1954).

4.4. Determination of nuclear electric quadrupole moments

It has just been mentioned in §4.3.2 that a nucleus having spin number greater than $\frac{1}{2}$ possesses in general an electric quadrupole moment, and that the interaction between this moment and its electric environment causes a displacement of the nuclear energy levels. As the fuller discussion in Chapter 8 shows, the displacements of the energy levels are functions of the product of the electric quadrupole moment Q and the electric field gradient at the nucleus. From measurements of this electric quadrupole splitting of the nuclear magnetic resonance spectrum for solid specimens the product of Q and the electric field gradient may be deduced. In order to obtain values of Q from this product it is necessary to calculate the gradient of the crystalline electric field, and up to the present this has not proved possible. Absolute values of Q are therefore not obtainable, and the method is limited to finding the ratio of the quadrupole moments of nuclei which are known to be subjected to the same electric environment. In practice one can only be certain that the environment is identical if one is dealing with two isotopes of the same element. Ratios of quadrupole moments have been obtained for ^{6}Li and ^{7}Li using a single crystal of spodumene $LiAl(SiO_3)_2$ (Schuster and Pake, 1951 a; Cranna, 1953), and for ^{63}Cu and ^{65}Cu using a single crystal of $K_3Cu(CN)_4$ (Becker and Krüger, 1951; Becker, 1951).

Perhaps the most important reason why this method has not been applied more widely is that in 1950 Dehmelt and Krüger (1950) developed the pure quadrupole resonance method, which yields the same information more directly and more accurately. A brief account is given of this method in §8.5.

Another source of information, though in this case very approximate, concerning the relative magnitudes of the quadrupole moments for an isotopic pair of nuclei, is the spin-lattice relaxation time T_1 for liquid specimens. If the interaction between the nuclear quadrupole moments and the fluctuating electric field gradient within the liquid is the dominant spin-lattice relaxation mechanism, then T_1 is proportional to Q^{-2} (see §8.2). Measured values of T_1 for both nuclei in the same liquid thus yield the ratio of quadrupole moments. The time T_1 is frequently short enough to be the factor which determines the width of the resonance line, and since in such cases the width is inversely proportional to T_1, a measurement of the relative widths of the two resonances gives the ratio of the relaxation times and thus in turn the ratio of the quadrupole moments. By this method Pound (1947 b, 1948 b) obtained approximate values for the ratio of the quadrupole moments of ^{79}Br and ^{81}Br using aqueous solutions of lithium and caesium

G

bromide, and for the ratio for ^{69}Ga and ^{71}Ga using an aqueous solution of gallic chloride. Similarly, Proctor and Yu (1951) using an aqueous solution of K_2MoO_4, were able to show that the quadrupole moment of ^{97}Mo exceeded that of ^{95}Mo.

We conclude this section with a note concerning higher order moments than the quadrupole moment. A nucleus whose spin number exceeds unity may also possess a magnetic octupole moment, while a nucleus whose spin number exceeds $\frac{3}{2}$ may possess an electric sedecipole moment. Using atomic beams the existence has been established of a nuclear magnetic octupole interaction in the $^2P_\frac{3}{2}$ state of iodine (isotope ^{127}I, spin number $\frac{5}{2}$) by Jaccarino, King, Satten and Stroke (1954), and in the $^2P_\frac{3}{2}$ state of indium (isotope ^{115}In, spin number $\frac{9}{2}$) by Kusch and Eck (1954). However, in his nuclear magnetic resonance studies of nuclear electric quadrupole interactions in crystals, Pound (1950) found no evidence of interaction between a possible nuclear electric sedecipole moment of ^{127}I with the fourth derivative of the electric potential in a single crystal of potassium iodide, and was able to give an upper limit for the interaction energy.

4.5. Determination of the proton magnetic moment in absolute units

It was mentioned in §4.1 that nuclear magnetic moments are usually measured in terms of the moment of the proton, and may be converted into other units in which the proton moment is known. In this section the measurement of the proton moment in absolute units is described, while the following two sections deal respectively with the measurement in nuclear magnetons and in Bohr magnetons. The importance of these measurements is not limited merely to the conversion of nuclear magnetic moments into useful units; as §4.8 will show, their interrelation leads to more accurate values for some of the fundamental physical constants. Furthermore, the measurement of the proton moment in absolute units provides a practical basis for the accurate measurement of magnetic fields (§4.9).

The measurement of the proton moment in absolute units, namely in erg gauss^{-1}, was carried out at the National Bureau of Standards in Washington, by Thomas, Driscoll and Hipple (1949,

1950 a, b). Their main task was the accurate absolute measurement of a magnetic field H_0 in which the proton resonance frequency ν_0 was determined. The resonance condition (4.1) may be rewritten as

$$\gamma \equiv \frac{\mu}{I\hbar} = \frac{2\pi\nu_0}{H_0} \; ; \qquad (4.6)$$

the ratio of frequency to field thus yields a value for the gyro-magnetic ratio γ. In order to convert this value into a value of the magnetic moment μ, a value for Planck's constant has to be taken from other work. (The spin number I is of course well known to be $\frac{1}{2}$ for the proton.)

The nuclear magnetic resonance system used the amplitude bridge of Thomas and Huntoon (1949), already described in §3.3, which is insensitive to microphonics and free from a dispersion component. The proton-bearing specimen consisted of 0·4 cm.[3] water containing enough dissolved ferric salt to reduce the relaxation time T_1 to such a value that the signal had a maximum amplitude without appreciable line broadening. The radiofrequency circuit was supplied from a 20 Mc./s. crystal oscillator whose frequency was calibrated against the Bureau's standard frequency station, and was known within one part in 10^6.

The resonance was found in a field of about 4700 gauss supplied by an electromagnet between approximately rectangular pole-faces 22 by 32 cm. The field was held constant by means of a proton regulator (see §4.10) within two parts in 10^6. With the radio-frequency system removed from the magnet gap, the strength of the field was determined by the vertical force exerted on a set of 10 cm. long copper wires carrying a measured current. The force was evaluated by comparison with the action of gravity on a standardized mass, which was placed on the scale pan of a balance when the current was reversed. The field measured was thus the mean value over the length of the 10 cm. conductors. It was therefore essential that it should be as homogeneous as possible, since the quantity one really wishes to measure is the field strength at the specimen. After shimming, the field was carefully mapped out with a proton sample (see §4.9), and a correction for the difference between the mean value and the local value was evaluated. The total uncertainty in determining the magnetic field H_0 amounted to 22 parts in 10^6, of which the principal contribution arose from the measurement of the dimensions of the conductors. By comparison the uncertainties in frequency measurement and in adjusting to the resonance condition were quite small.

The use of a specimen containing dissolved paramagnetic salt left an uncertainty in the results in view of the difficulty in making a correction for the paramagnetic effect (see §4.1). Fortunately it was found that substitution of an oil sample indicated no observable shift of the resonance, and subsequently Thomas (1950) directly compared the resonance condition for this oil specimen with that for hydrogen gas. Taking into account also the calculated value of Ramsey (1950 b) for the diamagnetic screening effect of the electrons for the hydrogen molecule (see §4.1), a correction of 28 parts in 10^6 had to be applied to the original result, giving for the proton gyromagnetic ratio the value

$$\gamma_p = (2\cdot675305_5 \pm 0\cdot00006) \times 10^4 \text{ sec.}^{-1} \text{ gauss}^{-1}. \tag{4.7}$$

4.6. Determination of the proton magnetic moment in nuclear magnetons

The method adopted for this determination consists in measuring in the same steady magnetic field H_0 both the proton magnetic resonance frequency ν_0, and also the frequency ν_c of orbital rotation of the proton, sometimes called the cyclotron resonance frequency. The latter frequency is well known to be

$$\nu_c = \frac{eH_0}{2\pi M_p c}, \tag{4.8}$$

where e is the charge on the proton in electrostatic units, M_p is the mass of the proton and c is velocity of light. From (4.1), for $I = \frac{1}{2}$, the proton magnetic resonance frequency is

$$\nu_0 = \frac{2\mu_p H_0}{h}, \tag{4.9}$$

where μ_p is the magnetic moment of the proton. By dividing equations (4.8) and (4.9) we get

$$\frac{\nu_0}{\nu_c} = \frac{\mu_p}{e\hbar/2M_p c} = \frac{\mu_p}{\mu_0}, \tag{4.10}$$

and we see that the ratio of the two frequencies is a direct measure of the proton magnetic moment in terms of the nuclear magneton μ_0.

The nuclear magnetic resonance frequency ν_0 is readily determined by the methods already described, while two pieces of apparatus have been developed independently for the measurement of ν_c: (i) the omegatron (Hipple, Sommer and Thomas, 1949; Sommer, Thomas and Hipple, 1950, 1951), and (ii) the decelerating cyclotron (Bloch and Jeffries, 1950; Jeffries, 1951).

The omegatron is so called because it is used, as also is the decelerating cyclotron, for the measurement of the angular frequency of orbital rotation of an ion, to which the symbol ω is frequently applied. In the omegatron protons are generated by

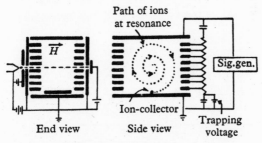

Fig. 22. Simplified diagram of the omegatron
(Sommer, Thomas and Hipple, 1951).

electron impact of residual hydrogen gas at the centre of a highly evacuated chamber placed between the poles of an electromagnet (fig. 22). A uniform radiofrequency electric field of the cyclotron resonance frequency is directed at right angles to the magnetic field and accelerates the protons; meanwhile a small positive potential applied to the guard rings prevents the protons escaping axially. When their orbits have reached a radius of 10 mm. the protons strike a collector whose current gives an indication of the resonance. Any ion having a different charge-to-mass ratio from that of the proton cannot reach the collector.

An analysis of the orbital motion (Sommer, Thomas and Hipple, 1951) shows that the sharpness of the resonance, as defined by the ratio of the resonance frequency to the full width of the resonance, is equal to $\frac{1}{2}\pi$ times the number of revolutions the protons make at resonance before reaching the collector. The radiofrequency field was therefore reduced to a low amplitude in order that the number of orbital revolutions might be as large as possible, and in practice was of the order of 10,000.

The omegatron and the proton resonance system could be quickly interchanged in the magnet gap and the resonant frequencies could thus be measured in essentially the same magnetic field. The final result was

$$\mu_p = (2\cdot79276 \pm 0\cdot00006)\mu_0. \tag{4.11}$$

The second method of determining the cyclotron resonance frequency (Bloch and Jeffries, 1950; Jeffries, 1951) did in fact use a cyclotron. It was, however, exceedingly small compared with conventional cyclotrons, since the dees were only 8·5 cm. in diameter, and could be placed in the very homogeneous region of the

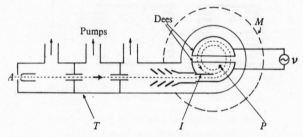

Fig. 23. Schematic diagram of the decelerating cyclotron (Bloch and Jeffries, 1950). A, arc source; I, injection plate; M, magnet poleface; P, collector probe; T, pumped tube.

field of a proton-regulated electromagnet producing about 5300 gauss between shimmed polefaces 26·4 cm. in diameter. As is shown in fig. 23, a beam of 20,000 eV. protons from the arc source A passed through a tube T about 1 m. long equipped with three differential pumping stages, and entered the cyclotron dee cavity near its periphery. When in the condition of resonance, the protons were decelerated along an inward spiral path until they reached the collector probe P, whose current was registered.

The decelerating mode of operation had the advantage that the proton beam could be injected from outside, and enabled a very low pressure to be maintained in the dee chamber independently of the arc pressure. In contrast with the conventional accelerating mode of operation, the decelerating cyclotron could be operated at odd harmonics of the proton cyclotron resonance frequency ν_c, and for two reasons this has the advantage of increasing the sharpness of the resonance. In the first place the use of the pth harmonic provides a correspondingly more accurate timing of the protons and reduces the resonance bandwidth by a factor p. Secondly, the transit time between the dees becomes a larger fraction of the radio-frequency period so that the mean decelerating force is reduced; this is especially true for the inner revolutions of the spiral. Consequently, with the dee radiofrequency voltage still high enough to allow clearance of the injection plate I after the first revolution, the number of revolutions necessary to reach the collector increases with p, giving an additional gain in resolution. The highest harmonic used was the 11th, the number of revolutions in the spiral orbit reaching about 500. The degree of precision of the final result was determined largely by the cyclotron resonance signal width, and since this was about three times the width obtained using the omegatron, the precision is correspondingly lower. The final result was

$$\mu_p = (2 \cdot 7924 \pm 0 \cdot 0002)\mu_0. \qquad (4.12)$$

Although the two results (4.11) and (4.12) differ by slightly more than their combined estimated errors, the difference cannot be regarded as significant, and indeed it is very satisfactory that the agreement is so good.

4.7. Determination of the proton magnetic moment in Bohr magnetons

For this determination (Gardner and Purcell, 1949; Gardner, 1951), measurements were made in the same steady magnetic field H_0, both of the proton magnetic resonance frequency ν_0 and also of the cyclotron frequency ν_e of the orbital rotation of the electron. The latter frequency is given by

$$\nu_e = \frac{eH_0}{2\pi M_e c}, \qquad (4.13)$$

which differs from (4.8) for the proton cyclotron frequency merely in the substitution of the electron mass M_e for the proton mass M_p. By dividing equations (4.9) and (4.13) we get

$$\frac{\nu_0}{\nu_e} = \frac{\mu_p}{e\hbar/2M_e c} = \frac{\mu_p}{\beta}, \qquad (4.14)$$

where β is the Bohr magneton. The ratio of the two frequencies is thus a direct measure of the proton magnetic moment in Bohr magnetons.

The nuclear magnetic resonance frequency ν_0 is readily determined by the methods already described. The electron cyclotron frequency ν_e is about 657 times as great as the proton magnetic resonance frequency ν_0, and thus falls in the microwave region for a steady field of several kilogauss. A field of about 3300 gauss was chosen, bringing ν_0 to about 14 Mc./s. and ν_e to about 9630 Mc./s., equivalent to a free-space wavelength of about 3·2 cm.

An evacuated rectangular waveguide (fig. 24), with its broad side parallel to the applied magnetic field H_0, was traversed by a

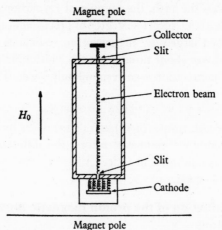

Fig. 24. Schematic diagram of electron resonator (Gardner and Purcell, 1949; Gardner, 1951). The proton resonance was obtained with a specimen placed alongside the waveguide.

beam of slow electrons from a hot cathode, which entered the guide through a narrow slit in the short side, and drifted across the guide in the direction of the magnetic field. The field prevented the ribbon-shaped beam from spreading, so that the electrons could pass through a similar slit in the opposite wall of the guide, where they were collected. When the guide was excited in the TE_{10}-mode at the cyclotron resonance frequency, the oscillatory electric field expanded the helical trajectories of the electrons; the resonance condition could then be recognized by the decrease in collector current resulting from the failure of the expanded beam to pass through the second slit. As with the experiments described in the preceding section, the sharpness of the resonance depended upon the number of revolutions in the helical path between the slits. This feature underlines the importance of the use of slow electrons; in practice for thermal electrons about 1000 revolutions were made.

It was found, however, that when the electron current was limited by a space-charge potential minimum within the guide, a much sharper *increase* of collector current appeared at resonance, provided that only a weak microwave field was applied. The expansion at resonance of the electron cloud at the potential minimum caused a reduction in depth of the minimum and hence an increase in collector current. The depth of the minimum is especially sensitive to the presence of electrons of very slow velocity, and these electrons make many more revolutions between the slits than the fast electrons. This velocity selection resulted in an improvement in resolution by a factor of 10 or so.

The proton resonance frequency was multiplied up 9 times, and was then fed into a crystal rectifier housed in a waveguide which generated a rich crop of harmonics. The 73rd harmonic, which was thus the 657th harmonic of the original frequency, was compared with the microwave electron resonance frequency using a spectrum analyser; the harmonic was identified with the aid of a cavity wavemeter. After applying a diamagnetic correction to the proton resonance frequency (§4.1), the final result was

$$\mu_p = (1 \cdot 52101 \pm 0 \cdot 00002) \times 10^{-3} \beta. \qquad (4.15)$$

Koenig, Prodell and Kusch (1952) have used the molecular beam method to find with high precision the ratio of the proton magnetic moment to the electron spin magnetic moment. They combine their result with that of the present section to find that the value of the electron spin magnetic moment is $(1 \cdot 001146 \pm 0 \cdot 000012)\beta$. This value is in excellent agreement with the theoretical value of Schwinger (1948) and Karplus and Kroll (1950), which differs from unity by the so-called radiative correction:

$$1 + \alpha/2\pi - 2 \cdot 973\alpha^2/\pi^2 = 1 \cdot 001145, \qquad (4.16)$$

where α is the fine-structure constant. These experiments thus provide support for the new methods of quantum electrodynamics on which the theoretical formula (4.16) is based.

4.8. Impact on the evaluation of the fundamental atomic constants

The experiments described in the three preceding sections are of such a precision that they provide valuable new input data for an evaluation of the fundamental atomic constants. A recent evaluation has been given by Dumond and Cohen (1953); here we merely seek to indicate the importance of the data by demonstrating some of the information which is directly obtainable.

First of all we notice that the experiments of §4.6 give μ_p/μ_0, while those of §4.7 give μ_p/β. The ratio of these two quantities is just the ratio of the Bohr and nuclear magnetons, which in turn is the ratio of proton and electron masses:

$$\left(\frac{\mu_p}{\mu_0}\right)\bigg/\left(\frac{\mu_p}{\beta}\right)=\left(\frac{\beta}{\mu_0}\right)=\left(\frac{e\hbar}{2M_ec}\right)\bigg/\left(\frac{e\hbar}{2M_pc}\right)=\frac{M_p}{M_e}. \qquad (4.17)$$

Secondly, we may combine the absolute measurement of the gyromagnetic ratio γ_p of the proton from §4.5 with μ_p/β from §4.7:

$$\frac{\gamma_p}{(\mu_p/\beta)}=\left(\frac{2\mu_p}{\hbar}\right)\left(\frac{e\hbar}{2\mu_pM_ec}\right)=\frac{e}{M_ec}. \qquad (4.18)$$

Since the electronic charge e is in electrostatic units, (4.18) thus gives the ratio $e_{\text{e.m.u.}}/M_e$ of the charge of the electron in electromagnetic units to its mass.

Similarly, we may combine the value of γ_p from §4.5 with μ_p/μ_0 from §4.6 to obtain the ratio of charge to mass for the proton:

$$\frac{\gamma_p}{(\mu_p/\mu_0)}=\left(\frac{2\mu_p}{\hbar}\right)\left(\frac{e\hbar}{2\mu_pM_pc}\right)=\frac{e}{M_pc}=\frac{e_{\text{e.m.u.}}}{M_p}. \qquad (4.19)$$

This last result when multiplied by the accurately known value of the isotopic weight of the proton, gives a value for the Faraday. This value, and also the values of M_p/M_e, $e_{\text{e.m.u.}}/M_e$, $e_{\text{e.m.u.}}/M_p$ from (4.17), (4.18) and (4.19) are obtained with an accuracy of about 30 parts per million.

Finally, the ratio μ_p/β may be combined with the measurement by Prodell and Kusch (1952) of the hydrogen hyperfine structure separation to give a value for the fine-structure constant α with an accuracy which is so good that uncertainties in the theoretical formula for the hyperfine structure separation become important.

4.9. Measurement of magnetic fields and related quantities

It was mentioned in §4.5 that the absolute determination of the proton gyromagnetic ratio provides a practical basis for the accurate measurement of steady magnetic fields. The method enables the measurement of magnetic fields to an accuracy of better than one part in 10^4 to be a straightforward proposition in any laboratory; this is a hundred-fold improvement over conventional methods. It is only necessary to measure the proton magnetic resonance frequency in the unknown magnetic field and to apply equation (4.6) to find its strength. In very precise work corrections must be made for the bulk diamagnetism (or paramagnetism) of the specimen and for the diamagnetic shielding of the electrons (§4.1).

The only limitation to the applicability of the method is that an observable signal-to-noise ratio be obtained. Generally a proton sample of water about 1 cm.^{3} is used, containing dissolved paramagnetic salt to shorten the relaxation time T_1; the density of protons is high, the intrinsic line width is narrow, and the proton gyromagnetic ratio is the highest of all the stable nuclei. All these factors aid the attainment of a good signal-to-noise ratio, as was discussed in §3.9.1. In that section it was also shown that the signal-to-noise ratio decreases rapidly with frequency, and therefore also with magnetic field strength for a given nuclear species.† On the other hand, there is no upper limit to the applicability of the method. At 10,000 gauss the proton resonance frequency of 42·6 Mc./s. begins to be rather inconveniently high (see, for example, Richter, Humphrey and Yost, 1954); for higher field strengths one may use a nucleus such as ^{7}Li having a lower gyromagnetic ratio than that of the proton, thus reducing the operating frequency approximately three-fold.

† Packard and Varian (1954) have, however, recently shown that by using a transient technique even the earth's magnetic field may be measured. A resultant magnetization was obtained of the protons in a large specimen of water by applying a field of about 100 gauss perpendicular to the earth's field. This field was then suddenly switched off, whereupon the resultant magnetic moment precessed about the earth's field. A receiving coil registered the nuclear induction signal caused by this free precession (see §5.5.2). The frequency of this signal, and therefore the earth's magnetic field also, could be measured with an accuracy of 1 in 10^4. Measurements of the earth's field strength with this precision may prove to be valuable in geophysical research and prospecting.

The most important requirement of the field to be measured is that it be fairly uniform. A very inhomogeneous field causes the resonance to be broad and therefore weak, and in any case the precision of the method, which is its main virtue compared with other conventional methods, cannot then be exploited. The field gradient must certainly be less than 100 gauss cm.$^{-1}$, and should preferably be less than 10 gauss cm.$^{-1}$.

Most of the systems described in Chapter 3 can be applied to the measurement of field strength; some of these have been developed as commercial instruments for the purpose. The marginal oscillator systems of §3.4 have a number of advantages over the other methods: they can be made compact and portable; they can cover a very wide frequency range by adjustment of the oscillator circuit only; they give a pure absorption signal free from dispersion component. An accurate frequency meter is required for the measurement of the resonance frequency. It is often necessary to make measurements at a number of places in the field, often in a rather confined space between magnet polefaces; for this reason the nuclear specimen and its radiofrequency coil are usually mounted at the end of a probe as with the conventional search coil. If the unknown field cannot be modulated at a low frequency, such modulation must be provided locally by supplying low-frequency current to a small pair of Helmholtz coils mounted at the end of the probe, with the specimen at the centre. The radiofrequency of the oscillator is adjusted until the modulated field sweeps through resonance, as indicated in the usual way on the oscillograph. The marginal oscillator systems of Hopkins (1949) and of Pound and Knight (1950) (see §3.4), were designed with a view to their use as magnetic field meters. Knoebel and Hahn (1951) avoid the complication of mounting Helmholtz coils on the probe by frequency modulating the radiofrequency instead of modulating the field. The frequency modulation is achieved by mechanical variation of a capacity in the oscillator tuned circuit, using an electrically driven reed. With this system the probe could be as narrow as 6 mm. diameter enabling precise measurements of magnetic field to be made in inaccessible places.

Since the proton magnetic moment has been measured absolutely with an accuracy of 22 parts per million (§4.5), magnetic fields may be measured with this order of accuracy. The relative values of field strength at two points may be found with still greater precision. Thus, the accuracy of determination of the ratio of two field strengths depends solely on the homogeneity of the two fields and the accuracy of frequency measurement, and in favourable cases may be carried out with an accuracy of one part per million. Moreover, if the same specimen is used throughout, no diamagnetic corrections are needed.

Small differences in magnetic field strength may be accurately resolved. The small variations in a fairly homogeneous field may be mapped out with an accuracy of one part per million; the magnet may then be shimmed and the nuclear magnetic resonance probe used to monitor the progress in improvement of uniformity. For this type of work it is desirable to use a very small specimen in order to obtain precise values of field at accurately located points.

The gradient of a magnetic field in a given direction may be obtained from relative measurements of the field at two adjacent positions along the line of interest using a small specimen. Alternatively, two small specimens may be investigated together, and the difference in their resonance frequencies, and hence of the field strengths, deduced from a transient beating effect between the two signals (see §5.5.1). A similar beating effect is obtained from a long specimen placed along the line of interest, which again enables the gradient to be calculated (§5.5.1). A rough estimate of the in-homogeneity of a magnetic field may be obtained from the breadth of the resonance signal for a liquid specimen, provided that one can assume, as is frequently the case, that the width of the signal is determined mainly by the spread of field strength over the specimen.

It has been pointed out by Béné, Denis and Extermann (1950a, 1951a, 1952a) that electric currents may be measured by passing them through coil systems free of ferromagnetic material, thus converting the current into a proportional magnetic field, and thence by nuclear magnetic resonance measurement to a proportional frequency. Since frequencies may be measured with great accuracy, a high relative accuracy of current measurement is obtained. In order to obtain absolute current values, the coil system may be calibrated either by calculation from its geometry, or by means of an accurately known current measured by a standard potentiometer method. A Helmholtz system of coils provides a very uniform field in which the nuclear magnetic resonance frequency may be measured. Since for convenient measurement the magnetic field must be at least several hundred gauss, the method is only applicable to the measurement of large electric currents.

The same workers (Béné, Denis and Extermann, 1950b, 1951b; Béné, 1951) were able to dispose of the suggestion of Stueckelberg (1948) that the resonance frequency in a magnetic field of given strength should differ according as the field is provided by a magnet or by an iron-free coil system.

4.10. Stabilization of electromagnets

Many types of apparatus incorporate an electromagnet whose field must be maintained constant over long periods of time. A conventional method of achieving this end employs an electronic device to maintain the magnet current constant.

The magnetic field is, however, not solely a function of the magnet current, but also depends upon the geometry of the magnet and the permeability of the iron, both of which may vary with the temperature. A greater constancy of magnetic field may therefore be obtained from a device which indicates changes of magnetic field, rather than changes of magnet current. Such a device may be based upon the nuclear magnetic resonance phenomenon.

A small specimen is chosen which gives a strong narrow resonance line; water containing dissolved paramagnetic salt is very satisfactory. The resonance is observed in the field to be controlled using one of the systems described in Chapter 3. If this field as a whole is not modulated at a low frequency, such modulation is provided locally in the manner described in §4.9. As fig. 25 shows, if the amplitude of field modulation is much less than the width of the resonance line, the resonance signal will be in the form of a low-frequency voltage whose amplitude is proportional to the first derivative of the resonance line. After passage through a phase-sensitive amplifier a d.c. signal is obtained proportional to this derivative. This derivative, also shown in fig. 25, is an ideal discriminator curve. When the field is adjusted for exact resonance, the derivative signal is zero. If the field drifts to either side, the derivative signal rises sharply and its sign is determined by the sense of drift. This signal may be amplified and used to control the magnet. If the magnet current is supplied from a motor generator, the control signal may be applied to the field windings of the

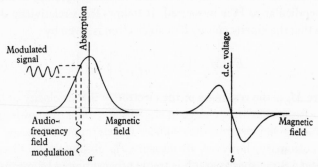

Fig. 25. (a) A typical absorption line shape. An audiofrequency magnetic field of small amplitude is superposed on the steady field, and is seen to result in an audiofrequency signal whose amplitude is proportional to the slope of the absorption line at the centre of the modulation sweep. (b) The first derivative of the absorption line to which the d.c. signal from the phase-sensitive amplifier is proportional.

generator; if the current is supplied from batteries, the control signal may be used to supply current to auxiliary coils on the magnet.

Proton-controlled magnetic field stabilizers have been described by Packard (1948); Thomas, Driscoll and Hipple (1950 b); Lindström (1951 b) and Thomas (1952). The magnetic field meters of Pound and Knight (1950) and Knoebel and Hahn (1951), discussed in §4.9, are also suitable when used in conjunction with a phase-sensitive amplifier. Stabilizers of this kind are now manufactured commercially. It is of course important to have a radiofrequency oscillator of high frequency stability. If the magnet is to be operated at one particular field strength, one may use a crystal-controlled oscillator of the appropriate frequency. The field can usually be held constant within a few parts per million.

4.11. Establishment of an absolute energy scale in β-ray spectroscopy

The work to be described in this section affords an excellent example of the application of nuclear magnetic resonance methods to the measurement of magnetic fields, and of the value of the greatly improved precision attainable.

The object of β-ray spectroscopy is to measure the kinetic energy spectrum of the β-ray emission from radioactive sources. The β-particles, or electrons, are subjected *in vacuo* to a uniform magnetic field $\mathbf{H}$, and the radius ρ of their circular trajectory in a plane perpendicular to $\mathbf{H}$ is measured. It follows from relativistic dynamics that the kinetic energy E of an electron is given by

$$E = M_e c^2 \left[\left\{ 1 + \left(\frac{H\rho e}{M_e c^2} \right)^2 \right\}^{\frac{1}{2}} - 1 \right], \qquad (4.20)$$

where M_e is the rest mass of the electron, c is the velocity of light, and e is the electronic charge in electrostatic units. A knowledge of $H\rho$ for a given electron thus suffices to determine its energy. It is not customary, however, to measure $H\rho$ absolutely in all cases; instead a β-ray spectrograph is frequently calibrated by means of a mono-energetic β-radiation whose $H\rho$ has been determined absolutely with as great precision as possible by other workers. Examples of such calibrating radiations are the K internal conversion electrons from (a) the 2·6 MeV. γ-ray of Th D, and (b) the 0·24 MeV. γ-ray of Th B. In the past the accuracy of determining the $H\rho$ value for such radiations has been limited by the accuracy with which the magnetic field could be measured; this was usually about one part in 10^3. Measuring the magnetic field with the aid of the proton magnetic resonance, Brown (1951) has determined $H\rho$ for example (a), and Lindström (1951 b, c, 1952) has determined $H\rho$ for both examples, all with an accuracy of about one part in 10^4; this is a ten-fold improvement in precision over earlier determinations.

Both workers used semicircular spectrographs. In Brown's case, the electrons traversed a semicircle of radius 11·4 cm. in a field of 880 gauss between the plane annular polefaces of a permanent magnet, and were registered on a photographic plate. Lindström used a proton-stabilized electromagnet with plane parallel polefaces 30 cm. in diameter, and the electrons were registered by two counters operating in coincidence mounted behind a narrow slit. The magnetic field was determined at points along the semicircular path of the electrons using a proton resonance probe as described in §4.9.

Strictly the experiments do not yield $H\rho$ directly but instead yield $v_0\rho$, where v_0 is the proton magnetic resonance frequency for the field strength to which ρ refers. The values could be converted into $H\rho$ values using the resonance relation $H = 2\pi v_0/\gamma_p$ and the absolute value of γ_p (§4.5). However, in order to calculate the kinetic energy of the electrons an alternative approach seems more attractive. We note that by algebraic rearrangement,

$$\frac{H\rho e}{M_e c^2} = \left(\frac{2\pi}{c}\right)\left(\frac{e\hbar}{2M_e c\mu_p}\right)\left(\frac{\mu_p H\rho}{\pi\hbar}\right). \tag{4.21}$$

Using (4.14), and the resonance relation just mentioned (with $\gamma_p = 2\mu_p/\hbar$), this gives

$$\frac{H\rho e}{M_e c^2} = \left(\frac{2\pi}{c}\right)\left(\frac{v_e}{v_0}\right)v_0\rho. \tag{4.22}$$

Thus the $v_0\rho$ values may be combined with the ratio (v_e/v_0) of the cyclotron frequency to the proton magnetic resonance frequency (§4.7) and with the velocity of light to give directly the dimensionless quantity on the left of (4.22), which enters into the energy formula (4.20). In order to evaluate the energy in electron volts, we require $Ec/10e$ where E is in ergs, c in cm.sec.$^{-1}$ and e in electrostatic units. The only additional quantities required are therefore the velocity of light and $(e/M_e c)$, the latter obtainable with precision from two other experiments involving the proton magnetic resonance, (§4.8, equation (4.18)).

In addition to determining the K conversion line from ThB, Lindström also measured the weaker L conversion line from ThB, and the K conversion line from the 0·51 MeV. γ-ray of ThC''. The difference in energy of the K and L conversion electrons of ThB is just the difference in energy between an electron in the K and L shells of the atom, and may be calculated from X-ray data. Since the difference in energy of the two radiations is thus known, and the ratio of their $H\rho$ values is measured as the ratio of the proton resonance frequencies for the two lines (using fixed ρ), it is therefore possible to calculate the energy and the $H\rho$ values of both radiations without knowing the field values in absolute units, and without measuring ρ (Siegbahn, 1944). The $H\rho$ values thus obtained may then be divided by the directly determined value of ρ to give absolute values of the magnetic field. These values in turn may then be taken with the directly measured proton resonance

H

frequencies to give an absolute value of the gyromagnetic ratio γ_p of the proton. The value of γ_p which Lindström obtained in this manner, namely, $(2 \cdot 67525 \pm 0 \cdot 00020) \times 10^4$ sec.$^{-1}$ gauss^{-1}, is in good agreement with that obtained by Thomas, Driscoll and Hipple (1949, 1950 a, b) (see §4.5), though its estimated error is about three times larger.

Lindström's $H\rho$ value for the K conversion line from the $0 \cdot 51$ MeV. γ-ray of Th C$''$ is in excellent agreement with that obtained by Hedgran (1951), who obtained his value from a comparison of the $H\rho$ values of the uranium external conversion electrons of (a) the Th C$''$ γ-ray and (b) annihilation radiation from ^{64}Cu. In calculating the annihilation energy it was assumed that the mass of the positron is equal to that of the electron. From the close agreement of the Th C$''$ $H\rho$ values Lindström concluded that the mass of the positron is indeed equal to that of the electron, within $0 \cdot 01 \%$.

CHAPTER 5

NUCLEAR MAGNETIC RESONANCE IN LIQUIDS AND GASES

5.1. Measurement of line width and spin-lattice relaxation time

Two of the most important characteristics of the nuclear magnetic resonance in any specimen are the spin-lattice relaxation time T_1, and the resonance line width, to which is closely related the spin-spin interaction time T_2 (these quantities were introduced in §§2.2 and 2.3). We shall therefore open this chapter with an account of methods for measuring these two quantities. We shall then give an account of the theory of nuclear relaxation in liquids, and go on to show that the theory is in good accord with the experimental facts. Our knowledge of relaxation effects in liquids is mainly due to the outstanding contributions of Bloembergen, Purcell and Pound (1947, 1948; see also Bloembergen, 1948), and this account is based upon their work.

5.1.1. Measurement of line width. The line width of a resonance absorption signal may be defined in a number of ways. For example, it may be defined as the interval between the two points at which the absorption component χ'' of the nuclear magnetic susceptibility falls to half its maximum value; alternatively, it may be defined as the interval between the points of maximum and minimum slope of the absorption curve; for solids the r.m.s. line width is most useful, as we shall see in Chapter 6.† It is not important which definition is taken provided it is used con-

† For Gaussian and Lorentz absorption line shapes the line width defined in the three different ways is as follows:

	(a)	(b)	(c)
Gaussian	$\left(\dfrac{\ln 2}{\pi}\right)^{\frac{1}{2}} \dfrac{1}{T_2}$	$\dfrac{1}{\sqrt{(2\pi)}\, T_2}$	$\dfrac{1}{\sqrt{(8\pi)}\, T_2}$
Lorentz	$\dfrac{1}{\pi T_2}$	$\dfrac{1}{\sqrt{3}\,\pi T_2}$	∞

where (a) is the line width measured between half-maximum points, (b) is that measured between the points of maximum and minimum slope, (c) is the r.m.s. line width defined by equation (2.60). T_2 is defined by equation (2.16).

sistently. The line width may be quoted either as a frequency interval $\delta\nu$, for which it is supposed that the magnetic field strength H_0 remains constant while the frequency is traversed through resonance, or else it may be quoted as a field interval $\delta H (= 2\pi\delta\nu/\gamma)$, for which it is supposed that the frequency remains constant while the field strength H_0 is traversed through resonance.

A strong resonance line may be displayed on an oscillograph (§3.1). If the sweep runs in synchronism with the modulation of the steady magnetic field, the trace provides a scale which is linear in field strength. The width of the line may then be determined from the oscillograph trace by direct measurement. The trace may be calibrated by observing the shift of a narrow resonance line brought about by a known change in radiofrequency or by a known change in magnetic field. Alternatively, the amplitude of modulation may be deduced from a measurement of the electromotive force induced in a small coil of known area-turns placed at the position of the specimen. If a correct measure of the line width is to be obtained three conditions must be satisfied. First of all the radiofrequency field strength H_1 must be small enough to prevent saturation, since, as equations (2.40) and (2.24) show, the absorption component χ'' of the nuclear magnetic susceptibility only varies with field or frequency in the same manner as does the line-shape function $g(\nu)$, when the saturation factor Z is effectively unity. Secondly, as was pointed out in §3.1, the resonance line must be traversed sufficiently slowly to avoid transient effects which modify the line shape. Thirdly, the amplifier must have an adequate bandwidth to pass the nuclear magnetic resonance signal without distortion.

For broader and weaker lines the phase-sensitive amplifier of §3.2 is used, and this yields the first derivative of the absorption line. Here also, it is important to avoid saturation. When using this method the definition of line width as the interval between maximum and minimum values of the derivative of the absorption line is clearly the most convenient. If the radiofrequency is traversed through the resonance condition, the values of frequency at the turning points may be measured directly with an accurate frequency meter. On the other hand, if the field is traversed through the resonance condition, usually by variation of current through auxiliary magnet windings, the values of current may be calibrated by measuring the frequency shifts of a sharp resonance line which accompany given changes of current in the field windings.

The line width obtained by both methods of measurement includes any broadening which may be caused by inhomogeneity of the steady magnetic field. In cases where the true line width is so narrow that it is masked by field inhomogeneity, the true value may nevertheless be obtained using methods based upon transient nuclear magnetic resonance effects, discussed in §5.5.

5.1.2. Measurement of spin-lattice relaxation time. There are three standard ways of measuring the spin-lattice relaxation time T_1: (1) the direct method, (2) the progressive saturation method, (3) transient methods. Since transient phenomena are considered later in §5.5, we shall defer consideration of the third group, and discuss only the first two methods in this section.

The direct method has the advantages of simplicity and straightforward interpretation. Using any of the experimental systems described in §§3.1, 3.2, 3.3 and 3.6, the resonance signal is first observed with a radiofrequency field H_1 small enough to avoid saturation. The value of H_1 is then increased to such a value that the saturation factor Z (see equation (2.24)) is reduced to about 10^{-1} or 10^{-2}; the susceptibility is in consequence reduced by the same factor (see (2.40)), while the spin temperature is raised to a value Z^{-1} times the lattice temperature (see (2.27)). The value of H_1 is then suddenly reduced to its original value; the observed signal then gradually returns to its original level according to the exponential relation (2.10), the characteristic time being just the spin-lattice relaxation time T_1 (equation (2.11)). A semi-logarithmic plot against time of the difference between the final signal strength and the strength at time t yields a straight line of slope $-1/T_1$, from which T_1 is found. Strictly, equation (2.10) governs the return to equilibrium in absence of a radiofrequency field, whereas the proper equation to apply here is (2.28), which is governed by a characteristic time $T_1 Z$. It is therefore important that during the return to equilibrium H_1 should be small enough to avoid appreciable saturation, so that the corresponding Z is effectively unity. Alternatively, the value of Z may be calculated and a correction made.

If the absorption signal is strong enough for display on an oscillograph, as is frequently the case with liquids, the re-growth of the signal may be observed visually provided T_1 is greater than a second or so. In practice it is rarely possible to make direct signal-versus-time measurements, since T_1 is usually less than 10 sec. for liquids. Instead, the recovery of the signal is photographed with a cine-camera operating at a known frame speed; the individual frames are then measured at leisure.

If the signal is weak, it is necessary to use the phase-sensitive amplifier and to record the signal strength with a meter. If T_1 is shorter than 10 sec. a high-speed recording meter is needed. It is of course clear that the time constant of the output circuit of the phase-sensitive amplifying system must be made much less than T_1.

When T_1 is too short for application of the direct method, the progressive saturation method may be used. It will be recalled that the saturation factor Z is a function of T_1 (equation (2.24)), and this allows T_1 to be derived from a study of the saturation process itself. Let us suppose that we are observing a signal proportional to the derivative $d\chi''/dH$ of the absorption curve, obtained with the usual field modulation system in conjunction with a phase-sensitive amplifier.

Using equation (2.40) for χ'', the derivative is

$$\frac{d\chi''}{dH} = \gamma \frac{d\chi''}{d\omega} = \tfrac{1}{2}\pi\chi_0\nu_0 \frac{d}{d\omega}[Zg(\omega)], \qquad (5.1)$$

where $\omega/2\pi$ is the radiofrequency. The output meter reading is not only proportional to $d\chi''/dH$, but also to the radiofrequency field H_1 and to the gain of the amplifier. It is arranged, however, that whenever H_1 is changed, the gain is changed in such a way as to keep the product of these latter two quantities constant; since the second detector current is an increasing function of the product of H_1 and radiofrequency gain, the required adjustment is readily made by altering the radiofrequency gain to keep the second detector current constant. The output meter reading $\varXi$ may then be written as

$$\varXi = \mathscr{C} \frac{d}{d\omega}[Zg(\omega)], \qquad (5.2)$$

where $\mathscr{C}$ is a constant. Two important cases may now be distinguished:

Case 1, for $\omega_m T_1 \ll 1$, where $\omega_m/2\pi$ is, as before, the modulation frequency; in this case T_1 is short compared with a modulation period, and in consequence the saturation factor Z varies during a cycle.

Case 2, for $\omega_m T_1 \gg 1$; in this case Z cannot change during a modulation cycle, and may be assigned a value appropriate to the centre of the modulation swing; being constant, the value of Z may then be taken out of the square bracket in (5.2) and placed in front of the differential operator.

As a guide to behaviour, let us suppose that the unsaturated absorption curve has the Lorentz or damped-oscillator shape given by Bloch's semi-macroscopic theory (§2.7). The form of $g(\omega)$ is therefore given from equation (2.59) as

$$g(\omega) = \frac{2T_2}{1 + (\omega - \omega_0)^2 T_2^2}. \tag{5.3}$$

For Case 1, we substitute in (5.2) for Z from (2.24) and for $g(\omega)$ from (5.3) and carry out the differentiation to find the meter reading $\mathscr{E}$. By differentiating again we find, by straightforward calculation that the maximum meter reading, $\mathscr{E}_{max}$, has the form

$$\mathscr{E}_{max} \propto (1 + \sigma)^{-3/2}, \tag{5.4}$$

where σ is defined as

$$\sigma = \gamma^2 H_1^2 T_1 T_2 = \frac{1}{Z_0} - 1, \tag{5.5}$$

using (2.25).

For Case 2, Z is placed in front of the differential operator in (5.2). We substitute for Z and $g(\omega)$ as for Case 1, and carry out the differentiation to obtain $\mathscr{E}$. Then by differentiating again, the maximum meter reading is found to have the rather complicated form

$$\mathscr{E}_{max} \propto [\{16 + 16\sigma + \sigma^2\}^{\frac{1}{2}} - 2 - \sigma]^{\frac{1}{2}} \times$$
$$[8 + 8\sigma - \sigma^2 + (\sigma + 2)\{16 + 16\sigma + \sigma^2\}^{\frac{1}{2}}]^{-1}. \tag{5.6}$$

Relative values of $\mathscr{E}_{max}$ are plotted against σ in fig. 26 for Cases 1 and 2. When the value of $\omega_m T_1$ lies between these two limiting cases, the $\mathscr{E}_{max}$ curve may be expected to lie between the two curves in fig. 26.

It has been implicitly assumed so far that the inhomogeneity of the steady magnetic field was negligible and does not contribute to the line width. When dealing with liquids this assumption is frequently not justified. We therefore consider now Case 3 for which the inhomogeneity width exceeds the true line width, which in turn

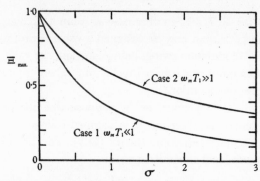

Fig. 26. Variation of maximum output meter reading $\Xi_{\max}$ with the saturation parameter σ defined as $\gamma^2 H_1^2 T_1 T_2$, for a homogeneous magnetic field.

we will suppose exceeds the field modulation amplitude. In this case the result is just an integration over all parts of the specimen, each of which is in a slightly different steady field, of curves of the kind analysed in Case 1 or Case 2 (depending on the magnitude of $\omega_m T_1$). In general, the decrease of $\Xi_{\max}$ with σ will follow much the same course as in Cases 1 and 2.

A rather difficult situation is sometimes encountered for liquids having relatively long relaxation times. Here the inhomogeneity of the steady field again frequently determines the line width; the true line width may, however, be very small and less than the field modulation amplitude. In this case it is the modulation amplitude H_m rather than the true line width which determines the degree of saturation in various parts of the specimen. In this situation, Case 4, Bloembergen (1948) shows that the degree of saturation is determined by the quantity $\gamma H_1^2 T_1 / H_m$ rather than $\gamma^2 H_1^2 T_1 T_2$.

In order to compare the saturation in various substances, or in one substance at various temperatures, the value of $\Xi_{\max}$ is plotted against log H_1, as H_1 is progressively increased from small values. The radiofrequency field strength H_1 is increased by increasing the signal generator output to which H_1 is proportional; one therefore actually plots the logarithm of the voltage output rather than log H_1. Fig. 27 shows some typical results of Bloembergen (1948) for water-glycerin mixtures of varying concentration. Theoretical curves for Cases 1 and 2 are also shown for comparison; the curves

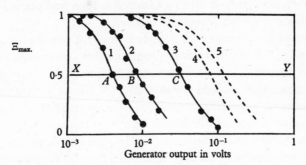

Fig. 27. Saturation curves for proton magnetic resonance absorption in mixtures of water and glycerin (Bloembergen, 1948). Curve 1 for 100% water, curve 2 for 40% water, curve 3 for 2% water. The broken lines 4 and 5 show for comparison the theoretical curves for Cases 1 and 2 (using the same value of $\gamma^2 T_1 T_2$ for both curves). The ordinate is the maximum output meter reading in arbitrary units.

for the two cases have very similar form and on a logarithmic plot differ mainly in a lateral shift corresponding to a factor of about 1·5 in H_1. Provided we are dealing with the same Case throughout, the relative values of Ξ_{max} are the same function of the saturation parameter σ (e.g. (5.4) or (5.6)). If therefore on a diagram such as fig. 27 we draw any horizontal line such as XY, then the points of intersection A, B, C with the experimental curves have the same value of σ. Hence for A and B we have

$$(\gamma^2 H_1^2 T_1 T_2)_A = (\gamma^2 H_1^2 T_1 T_2)_B, \tag{5.7}$$

and we find for the ratio of spin-lattice relaxation times

$$\frac{T_{1_A}}{T_{1_B}} = \left(\frac{H_{1_B}}{H_{1_A}}\right)^2 \frac{T_{2_B}}{T_{2_A}}, \tag{5.8}$$

where (H_{1_B}/H_{1_A}) is given by the antilog of AB in fig. 27. Provided that the line shapes are geometrically similar, the ratio (T_{2_B}/T_{2_A}) is just the ratio $(\delta H_A/\delta H_B)$ of the line widths, as is clear from the form of definition of T_2 (equations (2.15) and (2.16)).

The progressive saturation method thus gives relative values of T_1. These may be converted into absolute values by calibration with a specimen whose T_1 is long enough to be measured by the direct method also. The accuracy is limited largely by the degree of validity of the two assumptions, (a) that we are dealing with the same Case throughout, and (b) that the resonance line retains the same shape; errors of up to a factor of 2 may well occur if these assumptions are not adequately satisfied.

The progressive saturation method would of course give *absolute* values of T_1 if the constant of proportionality between H_1 and the signal generator voltage were known accurately. It is not difficult to estimate this constant roughly from the circuit parameters, and the order of magnitude knowledge of the values of H_1 thus obtainable is often valuable. Although no attempt appears to have been made, this apparatus constant could no doubt be determined more accurately, enabling accurate absolute values of T_1 to be obtained without the need for calibration by the direct method.

Suryan (1951) has suggested a method for measuring T_1 for a liquid by making the liquid flow through a tube around which the specimen coil is wound. The nuclei enter the coil in an unsaturated condition and are progressively saturated as they proceed through the coil. The signal strength is therefore greater than would be the case if the liquid had been stationary and subjected to the radio-frequency field for a long time. Making certain approximations it is found that the fractional increase in signal strength resulting from the flow is given by vT_1/l, where v is the velocity of liquid flow, and l is the coil length; T_1 can thus be found. A continuous-flow system appears to offer a method of improving the signal-to-noise ratio for weak signals. The method has been explored by Denis, Béné and Extermann (1952), Bloom and Shoolery (1953) and Sherman (1954).

5.2. Theory of spin-lattice relaxation in liquids

As we saw in the preliminary discussion in §2.2, spin-lattice relaxation is a process in which energy is exchanged between the nuclear-spin system and the lattice, tending to bring the two systems to a common temperature. The approach to equilibrium is exponential, the time constant of the process, T_1, being inversely proportional to the probability of transitions between the nuclear magnetic energy levels brought about by interaction with the lattice. Any theory of spin-lattice relaxation must therefore discover the important mechanisms for producing these transitions.

For non-metallic liquids† the important mechanism is that arising from the thermal motion of the atoms or molecules which constitute the 'lattice'. The atoms or molecules are regarded as vehicles conveying the nuclei from point to point. Thus each nuclear magnetic moment takes part in the random translational and rotational Brownian motion of the molecules. In consequence, the local magnetic field at any point contributed by the neighbouring nuclear magnetic moments, and by any electronic magnetic moments which may also be present, is a rapidly fluctuating function of time. The component at the resonance frequency ν_0 of the Fourier spectrum of this fluctuating field induces transitions between the nuclear magnetic energy levels in just the same way as does an applied radiofrequency field.

In Appendix 3 the probability of transitions induced in this manner is calculated. First a formulation is made of the energy of magnetic interaction between the ith nucleus and all the neighbouring magnetic moments, of which the jth is typical. For each neighbour, the energy is found to involve three functions of the distance r_{ij} between the nucleus i and the neighbour j, the angle θ_{ij} between $\mathbf{r}_{ij}$ and $\mathbf{H}_0$, and the azimuth angle ϕ_{ij} of $\mathbf{r}_{ij}$ measured from some fixed reference direction. The three functions are:

$$\left.\begin{aligned} Y_{0j} &= (1 - 3\cos^2\theta_{ij})r_{ij}^{-3}, \\ Y_{1j} &= \sin\theta_{ij}\cos\theta_{ij}\exp(i\phi_{ij})r_{ij}^{-3}, \\ Y_{2j} &= \sin^2\theta_{ij}\exp(2i\phi_{ij})r_{ij}^{-3}. \end{aligned}\right\} \qquad (5.9)$$

The intensities of the Fourier spectra of these three position functions are then introduced and are given the symbols $J_{0j}(\nu)$, $J_{1j}(\nu)$, $J_{2j}(\nu)$ respectively at frequency ν. The total intensity for all neighbours of the functions Y_{0j} is $\sum_j J_{0j}(\nu)$, and this is called $J_0(\nu)$; similarly, $J_1(\nu)$ and $J_2(\nu)$ represent the total intensities for the other two functions. It is found that fluctuations of the position coordinates are capable of inducing transitions between the energy levels of the ith nucleus for which the magnetic quantum number m_i changes by unity. Pictorially we may regard the motion of neighbour j with frequency ν_0 as producing at nucleus i an oscil-

† Liquid metals are discussed in §7.5.

lating magnetic field of this frequency which is capable of inducing transitions. It is further found that fluctuations of the position functions Y_{2j} at frequency $2\nu_0$ are capable of inducing transitions. This double-frequency effect arises as follows. The precessing components of the magnetic moment of j produce right- and left-hand circularly polarized local fields at i, one of which has the correct sense to cause transitions in nucleus i; this is the spin-spin mutual energy exchange process referred to in §2.3, which involves no exchange of energy between spin system and lattice. The other circularly polarized component of opposite sense has a negligible effect in a rigid lattice. If, however, molecular fluctuation imparts to the neighbour j a suitable motion, at frequency $2\nu_0$, the sense of rotation may be reversed, allowing the originally ineffective component to interact with nucleus i. Strictly speaking the frequency is only $2\nu_0$ when the neighbour j is identical with nucleus i. In the more general case when i and j are different, the frequency concerned is the sum of the precession frequencies for i and j. Fluctuations of the remaining position functions Y_{0j} do not induce transitions.

After calculating the total probability W of transitions induced by molecular fluctuation, the spin-lattice relaxation time T_1 is found using (2.11). If all the nuclear magnetic moments are identical, and no electronic magnetic moments are present, T_1 is found to be given (Appendix 3, equation (25)) by

$$\frac{1}{T_1} = 2W = \tfrac{3}{2}\gamma^4\hbar^2 I(I+1)[J_1(\nu_0) + \tfrac{1}{2}J_2(2\nu_0)], \qquad (5.10)$$

where, as usual, γ and I are respectively the nuclear gyromagnetic ratio and the nuclear spin number. As is shown in Appendix 3, if other magnetic moments, nuclear or electronic, are also present, $1/T_1$ is given by the sum over all types of magnetic moment of expressions similar to (5.10).

Although expression (5.10) is given for general values of I, it has to be remembered that nuclei having $I > \tfrac{1}{2}$ possess also an electric quadrupole moment. The interaction between this quadrupole moment and the fluctuating gradient of the local electric field constitutes another mechanism for spin-lattice relaxation in liquids. We shall, however, postpone consideration of this contribution until §8.2.

It is next necessary to find the form of the spectral intensity functions $J(\nu)$. We introduce a correlation time τ_c, which sets a time scale to the random motion. Roughly speaking τ_c is of the order of the time a molecule takes to turn through a radian, or to move through a distance comparable with its dimensions. It is found (Appendix 3, equation (15)) for the position function Y_{0j} that

$$J_0(\nu) = \sum_j J_{0j}(\nu) = \sum_j \overline{\mid Y_{0j} \mid^2} \, 2\tau_{cj} (1 + 4\pi^2 \nu^2 \tau_{cj}^2)^{-1}, \qquad (5.11)$$

where $\overline{\mid Y_{0j} \mid^2}$ is the time average of $\mid Y_{0j} \mid^2$. In general, the correlation times τ_{cj} for the different neighbours are not the same. Similar expressions to (5.11) for $J_1(\nu)$ and $J_2(\nu)$ are obtained by changing throughout the subscript 0 to 1 and 2 respectively.

In order to evaluate T_1 for a given liquid it is now necessary to calculate the $J(\nu)$ from (5.11) and substitute in (5.10). As an example the theory is applied to water. A given proton has as its nearest neighbour the other proton in the H_2O molecule, and we consider first the effect of this important intramolecular† neighbour. The molecule is regarded as rigid, and the orientation of the vector joining the two protons is assumed to vary randomly, no direction being preferred. Only the angular coordinates vary with time, and the functions Y_1 and Y_2 are given by

$$\left.\begin{aligned} Y_1 &= \cos\theta \sin\theta \, e^{i\phi} r_0^{-3}, \\ Y_2 &= \sin^2\theta \, e^{2i\phi} r_0^{-3}, \end{aligned}\right\} \qquad (5.12)$$

where r_0 is the interproton distance. (The subscripts i and j are omitted since there is but one neighbour under consideration.) The time averages of the $\mid Y \mid^2$ are obtained by averaging (5.12) over all spatial directions, and are found by straightforward integrations to be

$$\overline{\mid Y_1 \mid^2} = \tfrac{2}{15} r_0^{-6}, \qquad \overline{\mid Y_2 \mid^2} = \tfrac{8}{15} r_0^{-6}. \qquad (5.13)$$

Substituting these values in (5.11) and in turn in (5.10), and remembering that $I = \tfrac{1}{2}$ for the proton, we find that the intramolecular contribution to $(1/T_1)$ is given by

$$\left(\frac{1}{T_1}\right)_{\text{intra}} = 2W_{\text{intra}} = \frac{3\gamma^4 \hbar^2}{10 r_0^6} \left[\frac{\tau_c}{1 + 4\pi^2 \nu_0^2 \tau_c^2} + \frac{2\tau_c}{1 + 16\pi^2 \nu_0^2 \tau_c^2} \right]. \qquad (5.14)$$

† In this chapter and the following chapter we shall frequently be applying the adjectives 'intramolecular' and 'intermolecular' to nuclear neighbours; the important distinction between these two rather similar words is readily remembered when one recalls that 'intra-' means 'within', and 'inter-' means 'between'.

The determination of the correlation time τ_c, required in (5.14), is closely related to the problem encountered in the theory of Debye (1945, Chapter V) of dielectric dispersion in polar liquids. Debye assumed as an approximation that the molecule could be treated as a sphere of radius a embedded in a viscous liquid, and found a correlation time given by

$$\tau_D = 4\pi\eta a^3/kT, \tag{5.15}$$

where η is the viscosity of the liquid, k is Boltzmann's constant and T is the absolute temperature. The function whose correlation time is required in the Debye theory is $\cos\theta$, the symmetry character of which is different from that of the functions with which we are concerned here. Because of this difference, the correlation time differs slightly, and is given by †

$$\tau_c = \frac{\tau_D}{3} = \frac{4\pi\eta a^3}{3kT}. \tag{5.16}$$

From measurements of the dielectric constant of water in the microwave region, Saxton (1946) has found the Debye correlation time τ_D at 20° C. to be $8\cdot1 \times 10^{-12}$ sec. Taken in conjunction with (5.15) this gives the quite reasonable value of $1\cdot4$ Å. for the effective radius a of the water molecule. Using (5.16) our correlation time τ_c is therefore $2\cdot7 \times 10^{-12}$ sec. Thus at all practical radiofrequencies $2\pi\nu_0\tau_c \ll 1$, and the term in the brackets in (5.14) reduces to $3\tau_c$. Taking for r_0 the approximate value of $1\cdot5$ Å., we find that $(1/T_1)_{\text{intra}}$ has the value $0\cdot12$ sec.$^{-1}$

We now consider the contribution to the spin-lattice relaxation process of the intermolecular protons. The fluctuation in the local field arising from the neighbours is mainly caused by their translational motion. Let us calculate the correlation time τ_{cj} for molecules in a spherical shell between radii r and $r+dr$ centred on nucleus i. One takes for τ_{cj} the time in which nucleus j moves a distance r in any direction relative to nucleus i. The relative motion of i and j is by diffusion, and may be described by means of the diffusion coefficient D of the liquid, which has the value $kT/6\pi\eta a$

---

† Bloembergen, Purcell and Pound (1948) point out that the angular factors of the functions Y_0, Y_1, Y_2 belong to the spherical harmonic $Y_2(\theta,\phi)$, whereas $\cos\theta$ belongs to $Y_1(\theta,\phi)$. The correlation time of $Y_l(\theta,\phi)$ for a sphere in a viscous liquid is $8\pi\eta a^3/l(l+1)kT$, which gives $\tau_c = \frac{1}{3}\tau_D$, where τ_c refers to $l=2$ and τ_D to $l=1$.

for spherical molecules of radius a. Remembering that the motion is three-dimensional, and that the molecules containing nuclei i and j both diffuse, we find from the well-known theory of Brownian motion (see Chandrasekhar, 1943, equations (173) and (174)) that

$$\tau_{cj} = r^2/12D = \pi\eta a r^2/2kT. \tag{5.17}$$

We use (5.10) to find the contribution to $(1/T_1)$ from this neighbour j; we put $I = \frac{1}{2}$ and use (5.11) to obtain the values of $J_{1j}(\nu)$ and $J_{2j}(\nu)$, and for the values of $\overline{|Y_{1j}|^2}$ and $\overline{|Y_{2j}|^2}$ we replace r_0 by r in (5.13). The contribution is then found to be

$$\tfrac{3}{10}\gamma^4\hbar^2 r^{-6}\left[\tau_{cj}(1 + 4\pi^2\nu_0^2\tau_{cj}^2)^{-1} + 2\tau_{cj}(1 + 16\pi^2\nu_0^2\tau_{cj}^2)^{-1}\right]. \tag{5.18}$$

In view of the factor r^{-6}, only close neighbours are important, and for these one finds, using (5.17), that $2\pi\nu_0\tau_{cj} \ll 1$; the square bracket in (5.18) thus reduces to $3\tau_{cj}$. Treating all the neighbours as independent (which is not quite true), we sum their contribution by integrating over a volume from infinity to the radius of closest approach, $r = 2a$. If N is the number of nuclei per cm.³, the intermolecular contribution to $(1/T_1)$ is thus found to be

$$\left(\frac{1}{T_1}\right)_{\text{inter}} = \int_\infty^{2a} \tfrac{3}{10}\gamma^4\hbar^2 r^{-6}\left(\frac{r^2}{4D}\right) 4\pi N r^2\, dr$$
$$= 9\pi^2\gamma^4\hbar^2\eta N/10kT.\dagger \tag{5.19}$$

Substituting numerical values in (5.19) for water at 20° C., we find $(1/T_1)_{\text{inter}} = 0.08$ sec.⁻¹. Taken with the intramolecular contribution of 0.12 sec.⁻¹, the total value of $(1/T_1)$ is 0.20 sec.⁻¹. The theoretical value of T_1 is thus 5.0 sec. The effect of the intramolecular proton neighbours thus appears to be rather more important than that of the intermolecular neighbours. The correlation time τ_{cj} for the intermolecular neighbours has, however, been rather crudely defined, so that the intermolecular contribution to $(1/T_1)$ might be in error by as much as a factor of 2.

The experimental value of T_1 for water free of dissolved oxygen is 3.6 sec. (Chiarotti and Giulotto, 1954). In view of the approx-

† Although the calculation is slightly different, this result is identical with that obtained by Bloembergen, Purcell and Pound (1948); our N, the number of nuclei per cm.³, is just twice their N_0, the number of molecules per cm.³ The theory of relaxation by translational diffusion has been extended by Torrey (1953).

imations made in the theory, the discrepancy between theory and experiment cannot be regarded as significant.

A more detailed comparison of theory and experiment is deferred until §5.4. We now turn to a theoretical consideration of the width of the resonance line in liquids.

5.3. Resonance line width for liquids

It was mentioned in §2.3 that an important source of line broadening is the local magnetic field produced by the neighbouring nuclear magnet. Classically the component of the local magnetic field parallel to the steady field $\mathbf{H}_0$ due to a dipole j of moment $\boldsymbol{\mu}$ aligned with $\mathbf{H}_0$ is $\mu(3\cos^2\theta_{ij}-1)r_{ij}^{-3}$, where $\mathbf{r}_{ij}$ is the vector joining the point concerned to the dipole j, and θ_{ij} is the angle between $\mathbf{r}_{ij}$ and $\mathbf{H}_0$. Since the magnitude of the magnetic moment of a nucleus is $\gamma\hbar[I(I+1)]^{\frac{1}{2}}$, the local magnetic field may therefore be expected to be of the order of

$$\gamma\hbar[I(I+1)]^{\frac{1}{2}}(3\cos^2\theta_{ij}-1)r_{ij}^{-3} = -\gamma\hbar[I(I+1)]^{\frac{1}{2}}Y_{0j}, \qquad (5.20)$$

using (5.9).

For a rigid arrangement of independently oriented identical nuclear neighbours we therefore expect the mean square local field to be

$$\gamma^2\hbar^2 I(I+1)\sum_j Y_{0j}^2. \qquad (5.21)$$

As we shall see in §6.2.3, when dealing with solids, the rigorously derived mean square width due to both the local field and also the spin-exchange process discussed in §2.3 is indeed given by (5.21) apart from a numerical factor $\frac{3}{4}$.†

With liquids, however, the local field is fluctuating, and the effect of the more rapidly varying Fourier components of Y_{0j} averages to zero; only the near-zero frequency components of the spectral density $J_0(\nu)$ contribute to the line width. Supplying to (5.21) the factor $\frac{3}{4}$ required by the more rigorous derivation, our argument leads to a mean square line width

$$\frac{3}{4}\gamma^2\hbar^2 I(I+1)\int_{-\nu_1}^{\nu_1} J_0(\nu)\,d\nu, \qquad (5.22)$$

† Strictly this statement is only correct if all the nuclei have an identical environment. In the present treatment of liquids we are assuming for simplicity that this is so.

where $\pm\nu_1$ are the limits of the frequencies regarded as being 'near-zero'.

The square root of (5.22) expresses the r.m.s. line width in units of magnetic field. Multiplication by a factor $\gamma/2\pi$ converts this square root into the r.m.s. line width expressed as a frequency $\delta\nu$. The limit of 'near-zero' frequencies, ν_1, should be of the order of the line width itself. We therefore take $\nu_1 = \xi\,\delta\nu$, where ξ is a numerical coefficient of the order of unity; suitable adjustment of this coefficient can, moreover, take up differences in the form of definition of line width (see §5.1). Putting this value of ν_1 in (5.22), we find that the line width $\delta\nu$ is given by the implicit equation

$$(\delta\nu)^2 = \frac{3}{16\pi^2}\,\gamma^4\hbar^2 I(I+1)\int_{-\xi\,\delta\nu}^{+\xi\,\delta\nu} J_0(\nu)\,d\nu. \tag{5.23}$$

This equation thus gives the contribution to the mean-square line width caused by the magnetic interaction of identical nuclear neighbours. If other magnetic moments are present another term must be added to the right-hand side of (5.23).†

The expression for $J_0(\nu)$ is given in (5.11). After substituting in (5.23) and carrying out the integration we get

$$(\delta\nu)^2 = \frac{3}{8\pi^3}\,\gamma^4\hbar^2 I(I+1)\sum_j |\,\overline{Y_{0j}}\,|^2 \tan^{-1}(2\pi\xi\,\delta\nu\,\tau_{cj}). \tag{5.24}$$

If all the correlation times τ_{cj} are extremely long, so that all $(2\pi\xi\,\delta\nu\,\tau_{cj})$ tend to infinity, the last factor in (5.24) has the constant value $\frac{1}{2}\pi$, and the expression reduces to

$$(\delta\nu)^2 = \frac{3}{16\pi^2}\,\gamma^4\hbar^2 I(I+1)\sum_j Y_{0j}^2, \tag{5.25}$$

which is just the expression for the mean-square line width for a rigid system of nuclei given by (5.21), after the numerical factor $\frac{3}{4}$ has been supplied and a factor $(\gamma/2\pi)^2$ applied to convert from field units to frequency units. On the other hand, if all the times τ_{cj} are extremely short, so that all $(2\pi\xi\,\delta\nu\,\tau_{cj})$ tend to zero, expression

† As is shown in §6.2.3, the contribution of non-identical neighbours of spin number I_f and gyromagnetic ratio γ_f to the mean-square width is obtained by adding to (5.21) a similar term for each neighbour in which $\gamma^2 I(I+1)$ is replaced by $\frac{4}{9}\gamma_f^2 I_f(I_f+1)$. The origin of the factor $\frac{4}{9}$ is discussed in §6.2.3 (see also Ayant, 1953). In consequence additional terms must be added to (5.22) and (5.23) in which the same replacement is made.

I

(5.24) just leads to a zero line width, $\delta\nu = 0$. For liquids the τ_{cj} are usually sufficiently short to make $(2\pi\xi\,\delta\nu\,\tau_{cj}) \ll 1$. Equation (5.24) then reduces to the explicit relation

$$\delta\nu = \frac{3\xi}{4\pi^2}\gamma^4\hbar^2 I(I+1)\sum_j \overline{|Y_{0j}|^2}\,\tau_{cj}. \tag{5.26}$$

From (5.16) and (5.17) we see that the τ_{cj} should be proportional to η/T. Since the viscosity η of a liquid usually decreases rapidly as the temperature T is raised, one expects the τ_{cj} and therefore also the line width $\delta\nu$ to decrease rapidly as the temperature is increased.

When the magnetic dipolar line broadening becomes small, other sources of broadening must be considered. Some of these were listed in §2.3, and of these we have already discussed the broadening effect of non-uniformity of the steady magnetic field, and the distortion of the true line shape by saturation, transient effects and inadequate amplifier bandwidth. The discussion of broadening caused by nuclear electric quadrupole interactions, which occur only for nuclei having spin number $I > \frac{1}{2}$, is deferred until §8.2. One other source of broadening which concerns us here is that arising from the limitation of the lifetime of a nucleus in a given state by spin-lattice relaxation. This lifetime is of the order of $2T_1$, since the transition probability W is given by $1/2T_1$ (equation (2.11)); in consequence the energy levels are spread over a range of order $\hbar/2T_1$, and this spreads the resonance frequency over a frequency range of the order of $1/(2\pi T_1)$. For liquids having a very short T_1, this source of broadening is frequently of roughly equal importance with the magnetic dipolar broadening. To show this let us consider the case of water. As a first approximation we neglect the contributions of the intermolecular neighbours in forming the sum in (5.26), and just consider the one intramolecular neighbour; on account of the r_{ij}^{-6} factor in Y_{0j}, this near neighbour is in fact the most important. Dropping the subscript j we have from (5.9)

$$Y_0 = (1 - 3\cos^2\theta)r_0^{-6}, \tag{5.27}$$

where r_0 is the interproton distance in the water molecule. The time average of $|Y_0|^2$ is found in the same way as in deriving (5.13) by averaging over all spatial directions. We find by a straightforward integration

$$\overline{|Y_0|^2} = \tfrac{4}{5}r_0^{-6}. \tag{5.28}$$

Substituting in (5.26), with $I = \frac{1}{2}$ for the proton, we find for the dipolar broadening

$$\delta\nu = \frac{9\xi}{20\pi^2} \gamma^4 \hbar^2 r_0^{-6} \tau_c. \tag{5.29}$$

From (5.14), remembering that $2\pi\nu_0\tau_c \ll 1$, the contribution to the line width caused by the limitation of lifetime by the intramolecular spin-lattice relaxation process is of the order

$$\frac{1}{2\pi T_1} = \frac{9}{20\pi} \gamma^4 \hbar^2 r_0^{-6} \tau_c. \tag{5.30}$$

Comparison of (5.29) and (5.30) shows that when τ_c is short enough for equations (5.26) and (5.29) to apply, the two contributions to the line width are indeed about equal. Physically this near equality results from the fact that the intensity of the local field spectrum, given by equations such as (5.11), is about the same at near-zero frequencies as it is at frequencies near ν_0, when $2\pi\nu_0\tau_c \ll 1$.

The line width may also be expressed in terms of the time T_2 introduced in §2.3. The near equality of the two line width contributions in the region of short correlation times implies that T_1 and T_2 are approximately equal in this region. For a hypothetical liquid which obeys the Bloch equations (§2.7), T_1 and T_2 are *exactly* equal in this region. In Bloch's theory T_1 and T_2 are introduced as relaxation times governing the return to equilibrium of the longitudinal and transverse components of the precessing magnetization vector. Both components are relaxed by the same fluctuating local magnetic field, but draw upon different sections of its frequency spectrum. However, when $2\pi\nu_0\tau_c \ll 1$, the correlation time is much shorter than the period of rotation of the transverse component, and the fact that this component is rotating is unimportant; both components therefore have the same relaxation time in this region.

5.4. Comparison of relaxation theory with experiment

In addition to providing the theory of relaxation effects which we have just discussed, Bloembergen, Purcell and Pound (1948) have also produced strong experimental support for it. We have already mentioned in §5.2 that the theoretical and experimental values of T_1 for water at 20° C. are in reasonable agreement. This in itself is valuable support, since postulated relaxation mechanisms are apt to produce values which disagree with experiment by many powers of ten; an example of such a discrepancy is given in §6.6. Nevertheless, a more searching test of the theory consists in finding whether T_1 has the dependence on the correlation time τ_c, which the theory predicts.

In this discussion the simplifying assumption will be made that for a given liquid at a specified temperature all the τ_{cj} are the same; the subscript j will therefore be dropped. This assumption is clearly incorrect for the more distant intermolecular neighbours, but their effect is so small anyway that this simplification will not be too gross. With this assumption the general expression (5.10) for T_1, coupled with expressions of the form of (5.11) for $J_1(\nu)$ and $J_2(\nu)$, gives

$$\frac{1}{T_1} = \frac{C_1\tau_c}{1 + 4\pi^2\nu_0^2\tau_c^2} + \frac{C_2\tau_c}{1 + 16\pi^2\nu_0^2\tau_c^2}, \qquad (5.31)$$

where

$$\left.\begin{array}{l} C_1 = 3\gamma^4\hbar^2 I(I+1)\sum_j \overline{|Y_{1j}|^2} \\ C_2 = \tfrac{3}{2}\gamma^4\hbar^2 I(I+1)\sum_j \overline{|Y_{2j}|^2}. \end{array}\right\} \qquad (5.32)$$

When $2\pi\nu_0\tau_c \ll 1$, as is often the case for liquids which are not exceptionally viscous, (5.31) becomes

$$\frac{1}{T_1} = (C_1 + C_2)\tau_c. \qquad (5.33)$$

Bloembergen, Purcell and Pound first measured the values of T_1 at 20° C. for a series of six hydrocarbons whose viscosity ranged from 0·005 to 2·6 poise. If for such a series the correlation time τ_c is mainly determined by the viscosity at a given temperature, as equations (5.16) and (5.17) suggest, it is to be expected from (5.33) that η and $1/T_1$ should be roughly proportional; this was in fact found to be the case.

In another experiment they measured T_1 for a series of water-glycerin solutions which cover a wide range of viscosity. The variation of T_1 with viscosity is shown in fig. 28. Exact proportionality between $1/T_1$ and η is not perhaps to be expected, since the progressive change of concentration must bring with it a change in the environment of each proton. Actually a 500-fold change in η is accompanied by a change of only 100-fold in T_1.

A better test was obtained in a third experiment in which the same substance was used throughout, and its viscosity was varied over a wide range by changing the temperature; in this case the

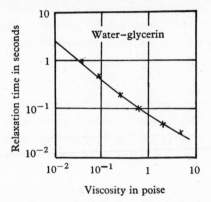

Fig. 28. Spin-lattice relaxation time T_1 for protons in water-glycerin solutions, measured at 29 Mc./s. (Bloembergen, Purcell and Pound, 1948).

environment of each proton should change very little. From (5.16) and (5.17) taken with (5.33) the theory leads us to expect in this case that $1/T_1$ should be proportional to η/T in the region of short relaxation times, where $2\pi\nu_0\tau_c \ll 1$, and should be independent of the radiofrequency ν_0, so long as the inequality is satisfied. Ethyl alcohol was the first substance used and the results are shown in fig. 29, where it is seen that within experimental error T_1 is inversely proportional to η/T, and is independent of ν_0.

The second substance used was pure glycerin. Here a very wide

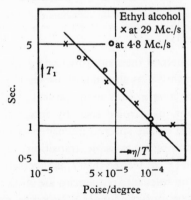

Fig. 29. Spin-lattice relaxation time T_1 for protons in ethyl alcohol, measured at 29 and 4·8 Mc./s. (Bloembergen, Purcell and Pound, 1948).

range (of order 10^4) in viscosity was obtained, since it was possible to supercool the liquid into an almost glass-like state. The approximation $2\pi\nu_0\tau_c \ll 1$ is no longer valid for the lower temperatures, and we must use the more general expression (5.31). For long correlation times, $2\pi\nu_0\tau_c \gg 1$, the expression becomes

$$T_1 = \frac{16\pi^2\nu_0^2\tau_c}{(4C_1 + C_2)}. \tag{5.34}$$

Thus at low temperatures where τ_c is long, we expect T_1 to approach proportionality with τ_c, whereas at higher temperatures where τ_c is short, we know from (5.33) that T_1 is inversely proportional to τ_c. Physically this dependence may be understood from fig. 30, where

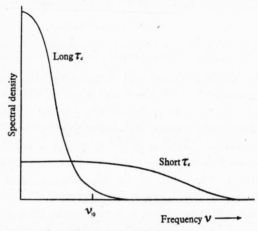

Fig. 30. Diagrammatic indication of the distribution of spectral density for long and short correlation times τ_c (i.e. $2\pi\nu_0\tau_c$ greater or less than unity, where ν_0 is the resonance frequency).

the form of the spectral intensity functions $J(\nu)$ from (5.11) for a single correlation time τ_c is shown both for short τ_c (high temperature) and also for long τ_c (low temperature). The area under the curves is constant, as can be seen by integrating (5.11); consequently for short τ_c, the spectral intensities at the frequencies ν_0 and $2\nu_0$ corresponding to possible transitions fall in proportion with τ_c, and from (5.10) we see that T_1 is inversely proportional to τ_c. For long τ_c the relevant frequencies are in the tail of the frequency distribution, and the longer τ_c becomes, the smaller is the spectral intensity at these frequencies; T_1 therefore increases with

τ_c. It follows that there must be a minimum value of T_1 in the intermediate region where $2\pi\nu_0\tau_c\sim 1$, and from (5.31) differentiation shows that the minimum lies in the range $\frac{1}{2} < 2\pi\nu_0\tau_c < 1$, the exact position depending on the relative values of C_1 and C_2 (defined in (5.32)). For the intramolecular contribution to $1/T_1$, we see from (5.14) that $C_2 = 2C_1$, and in this case one finds that the minimum occurs at $2\pi\nu_0\tau_c = 1/\sqrt{2}$.

The experimental values of T_1 for glycerin taken at two different radiofrequencies are plotted in fig. 31 against η/T, to which τ_c

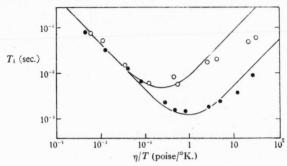

Fig. 31. The spin-lattice relaxation time T_1 for protons in glycerin plotted against the ratio of viscosity to absolute temperature (Bloembergen, Purcell and Pound, 1948). The open circles refer to measurements at 29 Mc./s. and the closed circles to measurements at 4·8 Mc./s.

should be proportional. Theoretical curves of the form (5.31) have been fitted to the points, taking $C_2 = 2C_1$ for simplicity. The logarithmic plot brings out the asymptotic dependences of T_1 upon τ_c in the regions of long and short τ_c. It will be noticed that in the high-temperature region (short τ_c) the values of T_1 are, as expected, independent of the radiofrequency ν_0. The expected minimum is displayed, though it is rather flatter than the theory predicts. The horizontal shift of the minimum with frequency is somewhat less than theory predicts. These discrepancies may partly be caused by experimental inaccuracy, but it must be remembered that the correlation times τ_{cj} have been assumed to be the same for all neighbours. A proper treatment introduces a distribution of relaxation times, which would have the effect of flattening the minimum in the theoretical T_1 curve. †

† Furthermore, Purcell (1953) has shown that the correlation function assumed in Appendix 3 (equation (13)) falls off too rapidly for a proper description of the intermolecular contribution; the correct function leads to a flatter minimum.

The line width in most of these experiments was very small and was in practice determined by the inhomogeneity of the magnetic field over the specimen. One exception to this statement was glycerin in its highly viscous supercooled condition, which reached a width rather more than 1 gauss at the lowest temperature of measurement. Making, as before, the simplifying assumption that the correlation times τ_{cj} are all equal, we find from (5.26) in the region of short τ_c, that

$$\delta\nu = C_3\tau_c \tag{5.35}$$

where $$C_3 = \frac{3\xi}{4\pi^2}\gamma^4\hbar^2 I(I+1) \sum_j \overline{|Y_{0j}|^2}. \tag{5.36}$$

The condition of shortness is less stringent than for the spin-lattice relaxation time approximations, because we merely require $2\pi\xi\,\delta\nu\,\tau_c \ll 1$, and for the glycerin measurements this should be satisfied at all except perhaps the very lowest temperatures. The experimental values of line width $\delta H\,(=2\pi\,\delta\nu/\gamma)$ are shown in fig. 32 plotted against η/T, to which τ_c should be proportional, and in fact the two quantities are found to be roughly proportional.

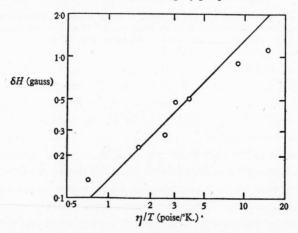

Fig. 32. Resonance line width δH for protons in glycerin plotted against the ratio of viscosity to temperature (Bloembergen, Purcell and Pound, 1948).

The effect of dissolved paramagnetic ions on the proton resonance in water has also been studied by Bloembergen, Purcell and Pound; the marked reduction in T_1 which results had already been demonstrated by Bloch, Hansen and Packard (1946 b). The magnetic moment of a paramagnetic ion is of the order of one Bohr magneton, and is thus some 10^3 times larger than a nuclear magnetic moment. The fluctuating local magnetic field will therefore be correspondingly larger and the relaxation time T_1 shorter. The relaxation produced by the ions is intermolecular, and we therefore seek an appropriate modification of expression (5.19), which referred to intermolecular relaxation by protons. In Appendix 3, it is pointed out that the modification is achieved by substituting μ_{ion}^2, where μ_{ion} is the effective magnetic moment of the ion, for $\gamma^2\hbar^2 I(I+1)$, which for the proton is $\frac{3}{4}\gamma^2\hbar^2$. The modified form of (5.19) is therefore

$$\frac{1}{T_1} = 12\pi^2\gamma^2\eta N_{ion}\mu_{ion}^2/5kT, \qquad (5.37)$$

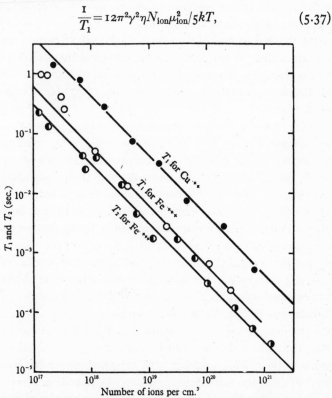

Fig. 33. Values of T_1 and T_2 for solutions of paramagnetic salts. ● T_1 for Cu^{++} ions, ○ T_1 for Fe^{+++} ions, ◑ T_2 for Fe^{+++} ions, all from Bloembergen, Purcell and Pound (1948). ◐ T_2 for Fe^{+++} ions from Gabillard (1952 b). Values of T_1 and T_2 for ferric ions, in excellent agreement with those shown, are reported by Chiarotti and Giulotto (1953) and Hahn (1950 b) respectively.

where N_{ion} is the number of ions per cm.³. This modification ignores any difference between the diffusion constant of the ion and that of a molecule of water. Equation (5.37) predicts: (a) that T_1 should be inversely proportional to concentration, and (b) that T_1 should be inversely proportional to μ_{ion}^2. The experimental results shown in fig. 33 confirm the first prediction, apart from some unexplained deviations at low concentrations. The second prediction has also been confirmed in a number of cases though there are discrepancies. Some of these may be due to the formation of complexes (Conger, 1953; Zimmerman, 1953, 1954). Other discrepancies arise from the fact that, as is pointed out in Appendix 3, in deriving (5.37) we continue to assume that the correlation time τ_c characterizing the fluctuating magnetic interaction is determined by the thermal motion of the ions in the liquid. In certain cases the relaxation time of the paramagnetic ion spins themselves may be shorter than the correlation time for thermal motion, and may then determine τ_c, leading to departures from (5.37) (cf. Conger and Selwood, 1952).

Apart from these exceptions, expression (5.37) gives a satisfactory quantitative value of T_1. For the Fe_{ion}^{+++} for example, the expression yields $T_1 = 0.08$ sec. for a concentration of 10^{18} ion cm.⁻³, taking $\eta = 0.01$ poise and $\mu_{ion} = 5.9$ Bohr magnetons; the experimental value at this concentration is 0.06 sec. It is to be noticed that quite a weak ionic concentration can have an important effect. Thus, a 0.01 molar solution of a ferric salt produces a reduction in T_1 from 3.6 sec. for pure water to 0.01 sec.

The line width in paramagnetic solutions is of special interest, since the extremely short spin-lattice relaxation time observed in concentrated solutions allows us to test the near equality of T_1 and T_2 discussed in §5.3. Values of T_2 are shown in fig. 33 for solutions of ferric ions, obtained by Bloembergen, Purcell and Pound (1948) from the observed line widths by assuming the frequency line width to be $1/(\pi T_2)$, together with values obtained by Gabillard (1952 b) using transient techniques described in the next section. It is observed that T_1 and T_2 are indeed roughly equal and have the same inverse dependence on ion concentration.

We thus conclude that the relaxation theory of Bloembergen, Purcell and Pound (1948) is well supported in its essentials by the results of the numerous experimental tests to which they and other workers have subjected it. As will be seen in later sections the theory has also been applied with success to gases and certain types of solid.

5.5. Transient effects

It was stressed earlier that if the true shape of the absorption or dispersion line is to be observed it is necessary to use a sufficiently low amplitude of radiofrequency field to avoid saturation, and to traverse the line sufficiently slowly to meet conditions (3.1) and (3.2). If these conditions are not met a wide variety of transient effects may be encountered, of which fig. 7 is but one example. Such transient effects are usually observed on an oscillograph screen, and in practice this means that a strong narrow signal is required. For this reason the study of transient effects has been mainly restricted to their observation with liquid specimens. In addition, when using pulse methods (§3.7), transient effects are necessarily met. Although not unrelated to the non-pulse methods, we shall for convenience treat the pulse methods separately.

5.5.1. Non-pulse methods. Six quantities having the dimensions of time enter into the description of the nuclear magnetic resonance process in the most general case: (1) the spin-lattice relaxation time T_1; (2) the spin-spin interaction time T_2, which is descriptive of the resonance line width; (3) the period $1/\nu_0$ of the radiofrequency field at resonance; (4) the period $2\pi/\omega_m$ of the modulation of the steady field; (5) the time actually spent in passing through the resonance line; (6) the time $(\gamma H_1)^{-1}$, which is a measure of the radiofrequency field strength. The pattern observed on the oscillograph at or near resonance depends upon the relative values of these six times. We shall not, however, consider all possibilities here, but shall instead devote our attention to several interesting cases. Some of the more general cases have been considered by Salpeter (1950) and by Gabillard (1951 b).

A rigorous analysis of transient effects would be extremely complicated, and does not appear to have been attempted. Instead, treatments are usually based upon Bloch's semi-quantitative theory (§2.7), which has the advantage of relative mathematical simplicity, and which suffices to show the main features. The results cannot of course be quantitatively exact, since, as was discussed in §2.7, such treatments necessarily assume the true absorption line to have a Lorentz shape, whereas actual line shapes only approximate to this shape.

We first discuss the transient effects observed by Bloch, Hansen and Packard (1946 b) in their pioneer experiments, and the analytical treatment of these effects by Bloch (1946); their apparatus has been described in §3.6. Although the experimental method has developed since these experiments were carried out, the early work is still very instructive. We consider the situation in which the relaxation times T_1 and T_2 are long compared with the time $(\gamma H_1)^{-1}$, and in which one passes through resonance in a time long compared with $(\gamma H_1)^{-1}$; this last condition often requires a relatively large value of H_1, and it was in fact the use of a large amplitude $2H_1$ of the radiofrequency field, of the order of 10 gauss, which characterized these early experiments. It is shown in Appendix 1 that with such conditions, and with the apparatus adjusted for reception of the dispersion component (u-mode), the nuclear magnetic resonance signal has amplitude proportional to

$$\frac{\mathscr{M}}{(1+\delta^2)^{\frac{1}{2}}}, \qquad (5.38)$$

where
$$\delta = (H_0 - \omega/\gamma)/H_1, \qquad (5.39)$$

and $\mathscr{M} = \mathscr{M}(t)$

$$= \int_{-\infty}^{t} dt' \, \frac{\mathscr{M}_0(t')\delta(t')}{T_1(1+\delta^2(t'))^{\frac{1}{2}}} \exp\left[-\int_{t'}^{t} \frac{\delta^2(t'') + T_1/T_2}{T_1(1+\delta^2(t''))} \, dt'' \right], \quad (5.40)$$

in which $\mathscr{M}_0$ is $\chi_0 H_0$, $\omega/2\pi$ is the radiofrequency and the quantity δ is a measure of the departure of the applied field H_0 from its resonant value. The main feature of the rather complicated expression (5.40) for $\mathscr{M}$ is that it can be either positive or negative, depending upon the positive or negative values which $\delta(t')$ in the integrand has assumed in the past.

Let us now suppose that the approach to resonance is sufficiently fast that the exponential in (5.40) does not appreciably change in the time during which δ changes by an amount of the order of unity. This is a case which Bloch calls 'rapid passage'. From the form of the exponential it is seen that this condition is better realized the larger are the times T_1 and T_2. One may then consider

$\mathcal{M}$ as practically constant in the region of resonance, its actual value depending mostly upon the values which $\mathcal{M}_0$ and δ assumed an appreciable time before the approach to resonance conditions. From (5.38) the resonance signal should then be proportional to $(1 + \delta^2)^{-\frac{1}{2}}$, having the form shown in fig. 34. The experiments

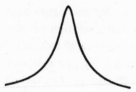

Fig. 34. The form of the dispersion signal (u-mode) under conditions of rapid passage (Bloch, 1946).

showed that if for a water specimen the field H_0 was set for some time at a value greater than the resonant value, and was then reduced through the resonance condition, the oscillograph trace did in fact have a form similar to fig. 34. If the field H_0 had initially been set below the resonant value and was then increased, the signal had the same shape, but was inverted. This inversion is in agreement with expectation, since the sign of $\mathcal{M}$ is determined by the sign of δ in the past, which is predominantly positive for decreasing fields, and negative for increasing fields.

In another experiment, using water, H_0 was set slightly above the resonant value, so that the resonant condition was just reached

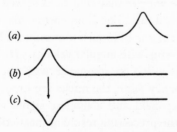

Fig. 35. Appearance of u-mode signal on oscillograph trace under conditions of rapid passage at three consecutive times as discussed in the text (Bloch, Hansen and Packard, 1946 b).

by the modulating field, and the signal appeared at the right-hand end of the oscillograph trace (Fig. 35a). The field H_0 was then suddenly reduced to a value rather less than the resonant value, so that the signal moved to the left-hand side of the trace (Fig. 35 b). Here the average value of δ is negative, and during a time of the order of T_1 the signal inverts (Fig. 35 c). From this observation it was correctly concluded that T_1 for water is a few seconds. With paraffin wax, however, the relaxation time was found to be much shorter, with the result that the signal inverted on its way across the trace. A concentrated solution of ferric nitrate had a still shorter relaxation time, so short in fact that the conditions of 'rapid passage' were not fulfilled; for this specimen T_1 and T_2 were both less than 10^{-4} sec., much shorter than the time required to traverse the resonance. The conditions were therefore those of 'slow passage', or effectively the steady-state conditions assumed in the earlier sections of this book. Under these conditions the trace should exhibit a typical dispersion curve similar to that shown in fig. 8, and this in fact was found to be the case. Drain (1949) and Gvozdover and Magazanik (1950) have discussed methods of measuring T_1 from experiments of this kind.

An interesting transient effect which is frequently encountered when using quite a small radiofrequency amplitude $2H_1$ is that exemplified in fig. 7. These 'wiggles', as Bloembergen, Purcell and Pound (1948) call them, occur *after* sweeping through a resonance line in a time short compared with the three times T_1, T_2 and $(\gamma \, \delta H_0)^{-1}$, where δH_0 is the inhomogeneity of the steady field H_0 over the specimen. These three times are those which determine the line width; the wiggles may thus be observed when conditions (3.1) and (3.2) are not met. The phenomenon may be pictured most readily from the nuclear induction viewpoint. The nuclei may be regarded as precessing with angular velocity γH_0; when H_0 reaches the value corresponding to resonance with the electromagnetic radiation of frequency $\omega/2\pi$, the nuclei are brought into phase by the radiofrequency field and a nuclear signal is induced in the receiving coil by the precessing resultant magnetic moment. As H_0 moves on beyond the resonant value, the resultant magnetic moment continues to precess with angular frequency γH_0, giving a nuclear induction signal of this frequency, which is no longer equal to ω. The nuclear induction signal therefore beats with the applied radiofrequency signal, giving alternate maxima and minima. As

Bloembergen (1948) points out, the phase difference between the nuclear induction signal and the applied radiation, after passing through resonance at time $t = 0$, is therefore given by

$$\phi(t) = \int_0^t \gamma[H_0(t) - H_0(0)]\,dt. \tag{5.41}$$

If H_0 changes linearly with time, as is approximately the case near the centre of a sinusoidal modulation sweep, then ϕ is given by

$$\phi = \tfrac{1}{2}\gamma\dot{H}_0 t^2, \tag{5.42}$$

and the beat signal should be of the form

$$\cos(\tfrac{1}{2}\gamma\dot{H}_0 t^2). \tag{5.43}$$

The beats persist so long as the nuclei remain in phase with each other and can thus form an appreciable resultant magnetic moment. In §2.3 we observed that phase coherence persists for a time of the order of T_2, and from the manner in which T_2 was introduced into the Bloch equations (see equations (2.48)), we would expect an exponential decay of the beat signal with characteristic time T_2. The beats should therefore have the form

$$\exp(-t/T_2)\cos(\tfrac{1}{2}\gamma\dot{H}_0 t^2). \tag{5.44}$$

A rigorous analysis by Jacobsohn and Wangsness (1948) based on the Bloch equations leads to signal shapes very similar in form to those given by (5.44), and in good agreement with experiment. The analysis by these workers shows that wiggles only occur when the sweep rate $\dot{H}_0$ exceeds $(4\gamma T_2^2)^{-1}$. The true line shape is therefore observed for

$$\dot{H}_0 \ll \gamma(\delta H_0)^2, \tag{5.45}$$

which is the requirement specified by (3.1) and (3.2).

Two other processes modify the decay of the beat signals. In the first place we notice from (5.44) that for a uniform field sweep, the frequency of the beats increases with time, and eventually passes beyond the pass-band of the amplifiers. The signal is then not registered on the oscillograph, even though the phase coherence may not have been fully destroyed. Secondly, if the true line width is very narrow, the observed resonance line width may be determined by the inhomogeneity width δH_0; in such cases $(\gamma\,\delta H_0)^{-1}$ is shorter than T_2. The decay is then predominantly caused by the interference between the beat patterns from different parts of the specimen, and $(\gamma\,\delta H_0)^{-1}$ determines the decay time rather than T_2.

Suppose the field is very homogeneous, so that the decay is determined by T_2; then if the period of field modulation $t_m\,(=2\pi/\omega_m)$ is less than, or of the order of, T_2, the beats have not decayed completely before the next passage through resonance at a time $\frac{1}{2}t_m$ later. As Gabillard (1951 c, 1952 a, b) has shown, the transients settle down to a steady state with the ratio of signal amplitudes just after and just before resonance equal to $\exp(t_m/2T_2)$. The oscillograph trace, using a linear timebase of period $\frac{1}{2}t_m$, should look somewhat like fig. 36 a. The ratio of signal amplitudes just after

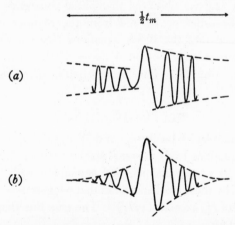

Fig. 36. Appearance of oscillograph trace under steady-state transient conditions: (a) in a perfectly homogeneous field, the trace envelope decays with characteristic time T_2; (b) in an inhomogeneous field, the envelope decays and recovers again in a time of the order $(\gamma\,\delta H_0)^{-1}$.

and just before resonance thus provides a means of determining T_2. The remarkable feature of this method of finding T_2 is that it works also when the field is so inhomogeneous that the decay is determined by $(\gamma\,\delta H_0)^{-1}$, rather than by T_2. In this case the trace resembles fig. 36 b. The signal decays rapidly in a time of the order of $(\gamma\,\delta H_0)^{-1}$. After the field sweep has made its full excursion and returns towards resonance, the signal builds up again, until just before resonance it has reached a fraction $\exp(-t_m/2T_2)$ of the value at the last resonance. The explanation of this decay and partial regrowth is as follows. Each small element of the specimen yields a decaying signal like fig. 36a; in an inhomogeneous field the signals from each element commence at slightly different times and therefore rapidly get out of phase as the field H_0 moves on from resonance, so that the resultant signal shows a rapid decay in a time of the order $(\gamma \delta H_0)^{-1}$. Within each element, however, the phases are still coherent, and are losing their phase relationship relatively slowly with the longer decay time T_2. As the field H_0 returns to resonance the phases return to coherence again and the signal grows once more. In fact, if there were no loss of phase coherence due to spin-spin and spin-lattice interaction the signal would increase to its initial value, giving a symmetrical decay and build-up. The essential features of the effect were first noticed by Gooden (1950), while the explanation was added by Gabillard (1951c, 1952a, b), who has used the effect to measure relatively long values of T_2 in the presence of an inhomogeneous field.

Let us now consider the shape of the envelope of decaying wiggles when the inhomogeneous field is the controlling factor. It turns out that the decay is not necessarily exponential in this case. Consider a specimen consisting of two small volume elements an appreciable distance apart in an inhomogeneous field. Each produces its wiggles, and the two patterns beat against each other, giving a resultant pattern like fig. 37.† In general, as Gabillard (1951c, d, 1952a, b) has shown, the beat envelope is given by the Fourier transform of the distribution function of the inhomogeneous magnetic field over the specimen. If we have a long specimen in a magnetic field having a uniform gradient, the distribution function is approximately square, being constant for field strengths between the values at the extremities of the specimen, and zero for other field values. This case, which is readily realized experimentally, also gives the 'beating of beats' pattern rather like fig.

† The same pattern is obtained in a homogeneous field with two specimens whose resonant frequencies are slightly different because they have a different chemical shift (Béné, Denis and Extermann, 1951e) or a different paramagnetic ion concentration (Gabillard and Soutif, 1951).

K

37.† Béné, Denis and Extermann (1953 a, b) have applied this effect to the measurement of small magnetic field gradients.

In addition to the work already mentioned, contributions to the subject of transients using non-pulse techniques have also been

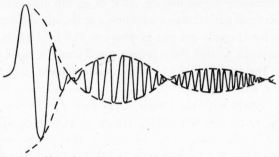

Fig. 37. 'Beating of beats' pattern obtained by interference of the 'wiggles' from two small volume elements situated at points where the field strength is slightly different.

made by Giulotto (1948); Bolle and Zanotelli (1948 b); Béné, Denis and Extermann (1950 c, d, 1951 b, c, d, 1952 b); Denis, Béné and Extermann (1952); Extermann, Denis and Béné (1949); Soutif and Gabillard (1951); Gabillard (1951 e); Soutif (1951); Ayant (1951 c); Bhar and Bhar (1952); Taylor (1953); Manus, Mercier, Denis, Béné and Extermann (1954).

5.5.2. Pulse methods. We mentioned in §3.7 that the pulse method developed by Torrey (1949) is concerned with transient behaviour during the time the radiofrequency pulses are applied, while the method of Hahn (1950 b) is concerned with the transient behaviour in the 'off' times between pulses.

We consider Torrey's method first. Neither the steady field H_0, nor the radiofrequency, are modulated; instead, the amplitude of the radiofrequency field is modulated with a sharp-fronted pulse, so that the approach to resonance takes place in a time short compared with the times $(\gamma H_1)^{-1}$, T_1 and T_2. The approach is thus more rapid than in Bloch's 'rapid passage', which was made in a time long compared with $(\gamma H_1)^{-1}$, though still short compared

† Actually the Fourier transform in this latter case is $(\sin \Delta t)/\Delta t$, where $\pm \Delta/\gamma$ is the range of field inhomogeneity, whereas in the former case of two elements in field strengths $H_0 \pm \Delta/\gamma$, the transform is $\cos \Delta t$; the envelopes are therefore only qualitatively similar in the two cases.

with T_1 and T_2. In order to obtain a simplified picture of what takes place, we first neglect relaxation effects. The equation of motion for the resultant magnetization vector $\mathcal{M}$ is then given by (2.44):

$$\dot{\mathcal{M}} = \gamma(\mathcal{M} \wedge \mathbf{H}), \qquad (5.46)$$

where $\mathbf{H}$ is the resultant applied magnetic field, which has a steady component $\mathbf{H}_0$ along the z axis and a component of amplitude H_1 rotating in the xy plane with angular velocity ω. We now transform to a system of axes rotating about the z axis† in the same sense as H_1 with angular velocity ω. If $\dot{\mathcal{M}}^*$ is the rate of change of $\mathcal{M}$ in the new system, then

$$\dot{\mathcal{M}}^* = \dot{\mathcal{M}} + \mathcal{M} \wedge \boldsymbol{\omega}. \qquad (5.47)$$

Using (5.46) this gives

$$\dot{\mathcal{M}}^* = \gamma(\mathcal{M} \wedge \mathbf{H}^*), \qquad (5.48)$$

where

$$\mathbf{H}^* = \mathbf{H} + \boldsymbol{\omega}/\gamma. \qquad (5.49)$$

In the rotating system $\mathbf{H}^*$ is constant; equation (5.48) thus corresponds to a precession of $\mathcal{M}$ about $\mathbf{H}^*$ with angular velocity

$$\begin{aligned} \omega^* &= \gamma \, | \, H^* \, | \\ &= [\gamma^2 H_1^2 + (\omega_0 - \omega)^2]^{\frac{1}{2}}, \end{aligned} \qquad (5.50)$$

using (5.49) and putting $\omega_0 = \gamma H_0$. Near resonance $\omega^* \ll \omega$, and in the laboratory frame of reference this motion appears as a slow nutation at frequency ω^* superimposed on the precession of $\mathcal{M}$ about $\mathbf{H}_0$ with frequency ω_0. The neglected relaxation processes damp the nutation. With the apparatus adjusted for observation of the absorption component (v-mode), a decaying oscillation is therefore observed during the on periods, the angular frequency of oscillation being ω^*. A full analysis by Torrey (1949), using the Bloch equations, of the case where $(\gamma H_1)^{-1} \ll T_1, T_2$, shows that the decay time near resonance is

$$2\left(\frac{1}{T_1} + \frac{1}{T_2}\right)^{-1}. \qquad (5.51)$$

---

† The use of rotating co-ordinates in magnetic resonance problems is discussed by Rabi, Ramsey and Schwinger (1954).

If the pulse duration is long compared with this time, the final signal is small, corresponding to saturation by the relatively intense radiofrequency field. The resultant magnetic moment of the specimen recovers exponentially during the 'off' time, with the characteristic recovery time T_1. Thus by observing the dependence of the initial amplitude on the duration of the 'off' time, the spin-lattice relaxation time may be measured. The spin-spin interaction time T_2 can then be found from the decay time of the nutational oscillations, given by (5.51). In many cases $T_1 \gg T_2$, so that a knowledge of T_1 is unnecessary for the evaluation of T_2 using (5.51).

We turn now to experiments in which one observes behaviour *after* the application of a radiofrequency pulse. If a short intense radiofrequency is applied to the specimen, a nutation is produced, as we have just shown, in the precession of the resultant nuclear magnetic moment vector $\mathcal{M}$ about H_0. In the rotating co-ordinate system this corresponds to a rotation of $\mathcal{M}$ about the field H^*, which at resonance is just H_1, and therefore is at right angles to H_0. At the end of the pulse the resultant magnetic moment vector $\mathcal{M}$ is therefore in general oriented in a new direction with respect to H_0. The component of $\mathcal{M}$ parallel to H_0 now differs from its equilibrium value. For $I = \frac{1}{2}$ this component is proportional to the excess n of nuclei per cm.³ in the lower energy state, and we see from (2.10) and (2.11) that this component will return to its equilibrium value exponentially with characteristic time T_1. The progress of this return to equilibrium may be inspected after any time t by applying an identical pulse to the first, the size of the received nuclear induction signal being proportional to n. From (2.10) and (2.11) we see that by plotting $\ln(1 - n/n_0)$ against t, where n_0 is the equilibrium value of n, T_1 is obtained from the slope of the resulting straight line. This method of measuring T_1, developed by Hahn (1949), is probably the most accurate method for the moderately short times encountered with liquid specimens; an accurate pulse-timing circuit is, of course, essential.

Immediately after the cessation of a radiofrequency pulse, the nuclear magnetic moments continue to precess coherently in phase for a time of the order of T_2. We encountered this effect in §5.5.1 where the beating between the nuclear induction signals of the freely precessing nuclei with the applied radiofrequency signal gave rise to 'wiggles' after passing through the condition of resonance. In the present case the applied radiofrequency signal is entirely cut off after the pulse, and in consequence the oscillograph trace merely registers the exponential fall, with characteristic time T_2, of the amplitude of the decaying nuclear induction signal caused by the free precession (Hahn, 1950a). As with the wiggles phenomenon, if the inhomogeneity of the field δH_0 is greater than the true resonance line width, the nuclear induction signals from various parts of the specimen interfere, and cause the signal to decay in the time $(\gamma \, \delta H_0)^{-1}$, rather than the longer time T_2. Nevertheless, in each small volume element of the specimen there is still phase coherence for a time T_2, and this is responsible for the remarkable effect of 'spin echoes', which we now discuss.

Hahn (1950 b) noticed that if two intense pulses of resonant radiofrequency power separated by a time interval t_0 were applied to a specimen, the oscillograph trace registered not only two nuclear induction pulses separated by a time t_0, but also a third spontaneous pulse at a time t_0 after the second pulse (see fig. 38).

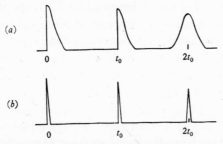

Fig. 38. Diagrammatic indication of an oscillograph trace in a spin-echo experiment. Radiofrequency pulses are applied at times $t = 0$ and $t = t_0$, and an echo appears at time $t = 2t_0$. The upper diagram illustrates the appearance when the steady magnetic field is fairly homogeneous over the specimen; the echo and the decay signals are broad. The lower diagram illustrates the appearance when the homogeneity is poor; the signals are sharp.

Hahn has called this spontaneous pulse a 'spin echo'. He has given an explanation of the effect by an analysis based on the Bloch equations (later extended by Das and Saha, 1954), and has applied the phenomenon to the measurement of T_1 and T_2.

We shall not give the rather complicated mathematical analysis here, but instead we shall give a simplified picture of the effect.

The first pulse at time $t=0$ changes the orientation of the resultant magnetic moment vector of each volume element of the specimen, and a nuclear induction signal is obtained from each as it precesses freely, decaying with a characteristic time T_2. Owing to the inhomogeneity of the magnetic field, the volume elements each have a different precessional frequency; the precessing magnetic moments and the nuclear induction signals they produce get rapidly out of phase in a time $(\gamma \delta H_0)^{-1}$ (which we take to be less than T_2), giving the first signal in fig. 38a. The second pulse at $t_0 < T_2$ may be regarded as reversing the procedure; the vectors are turned round again and while precessing pick up their phase differences at the same rate as previously they had lost them, and after another time t_0, find themselves again in phase so that a nuclear induction signal is reconstituted.† The echo signal thus produced is only fleeting, since the recovered phase coherence is only temporary, and at times after $2t_0$ the phases disperse once more. There is no essential mystery about the appearance of the echo signal at time $2t_0$, since after all, if the field had been perfectly homogeneous the observed signals would never have decayed between the pulses, except of course at the slower rate determined by T_2. If the magnetic field H_0 is very inhomogeneous, the loss of phase coherence is much more rapid, giving sharp signals at times 0 and t_0 (fig. 38 b), and the echo signal is equally sharp at $2t_0$, since the duration of recovered phase coherence is much shorter; in fact all three signals have a width of the order $(\gamma \delta H_0)^{-1}$.

The size of the echo signal at time $2t_0$ is determined by the exponential decay by a factor $\exp(-2t_0/T_2)$, of the nuclear induction signal of each small volume element. Hence by plotting the logarithm of the echo amplitude against various values of $2t_0$, one obtains T_2 from the slope of the resulting straight line.

The recovery of phase coherence at time $2t_0$ brings about a condition of the specimen comparable with that which followed the first pulse. Consequently if a third pulse is applied at a time t_1

† In a popular article Hahn (1953) likens the precessing nuclei to athletes on a running track. When the starter fires his gun (first pulse) the runners set off together but the field soon spreads out (decay of initial nuclear induction signal). The starter then fires again (second pulse); the runners all turn about and, running at the same rate, close up again and reach the starting post together (spin echo).

(i.e. at a time $(t_1 - 2t_0)$ after the echo), yet another echo should appear after a further interval $(t_1 - 2t_0)$, namely, at time $(2t_1 - 2t_0)$. Such secondary echoes are in fact observed, and are also observed at times $(t_1 + t_0)$, $(2t_1 - t_0)$, $2t_1$ making seven observed signals in all. The signal at time $(t_1 + t_0)$ is of particular interest, since, as Hahn's analysis shows, the size of this echo falls as $\exp(-t_1/T_1)$, thus allowing T_1 to be found.

The technique of spin echoes has been recently extended by Carr and Purcell (1954) to give more convenient ways of measuring T_1 and T_2, and to give a direct measurement of self-diffusion coefficients in suitable liquids.

Other contributions to the study of transients using pulse techniques have been made by Bradford et al. (1951 a, b), Clay et al. (1951), Strick et al. (1951), and Torrey (1952).

5.6. The chemical shift

Knight (1949) first showed that the resonance frequency for a given nucleus in a given field was sometimes dependent upon the chemical form in which the element was present. The main effect which he found was peculiar to metallic specimens and its discussion will be deferred until §7.3. Knight (1949) also noted a smaller difference in resonance frequency among some non-metallic phosphorus compounds, and shortly afterwards Dickinson (1950a, 1951) and Proctor and Yu (1950c, 1951) found further evidence of this frequency difference using non-metallic liquid specimens containing ^{14}N and ^{19}F nuclei in different chemical compounds. This difference in the resonance condition is called the 'chemical shift', and has now been found for a number of elements. Usually the discrepancy is only of the order of one part in 10^3 or 10^4, though for ^{59}Co Proctor and Yu (1951) find a difference of $1\cdot3\%$ between resonance frequencies in aqueous solutions of $K_3Co(CN)_6$ and $K_3Co(C_2O_4)_3$.

As we pointed out in §4.1 the existence of this shift sets a limit to the precision with which nuclear magnetic moments may be measured. In §4.1 it was explained that the shift very probably arises from a second-order paramagnetic term in the magnetic shielding correction of the electronic system of the molecule containing the nucleus, the shift depending upon the chemical compound, since the electronic configuration differs with the form of chemical binding. Although the theory has been given by Ramsey (1950a, b, 1951, 1952 b), the shift has proved calculable only in the case of molecular hydrogen. Nevertheless, several qualitative features support Ramsey's explanation.

First of all, if the shift of resonance frequency originates in the electronic shielding correction it should be proportional to the steady field strength H_0; this was found to be the case by Dickinson (1951) and Proctor and Yu (1950c, 1951). Secondly, if the discrepancy arises from a second-order paramagnetic term, it should be found only for molecular compounds, and not found when the element is present in the form of an ion or a free atom. This accords with the finding of Dickinson (1951) that there are no measurable shifts for ^{7}Li, ^{23}Na and ^{27}Al in aqueous solutions of salts of these elements; in each of these cases the atoms in solution should be almost completely ionized. Finally, the large chemical effect shown by ^{59}Co implies, in terms of Ramsey's explanation, the existence of excited electronic energy levels relatively close to the ground state; there should therefore be a slight temperature-dependence of the shift, and this has been observed by Proctor and Yu (1951).

Although a theoretical value of the shift cannot in general be calculated, it is clear that the shift should depend on the form of chemical binding of the atom concerned. Chemists have in fact made some headway in correlating the observed shifts with the electronic structures of molecules and the nature of their chemical binding (Gutowsky and Hoffman, 1950, 1951; Liddel and Ramsey, 1951; Arnold and Packard, 1951; Gutowsky, McCall, McGarvey and Meyer, 1951, 1952; Meyer and Gutowsky, 1953; Meyer, Saika and Gutowsky, 1953; Gutowsky and Saika, 1953; Gutowsky and McGarvey, 1953a, b; Shoolery, 1953; Masuda and Kanda, 1953, 1954; Saika and Slichter, 1954; Gutowsky and McCall, 1954; Ogg, 1954).

When a molecule possesses several resonant nuclei in different electronic environments, each nucleus has a different chemical shift and hence a different resonant frequency. With a liquid specimen in a highly uniform magnetic field the resonance therefore shows a fine structure. A good example, due to Arnold, Dharmatti and Packard (1951), is shown in fig. 39. Here the specimen was ethyl alcohol, and the three lines correspond, reading from the left, to the protons in the CH_3, CH_2 and OH groups respectively. This identification is achieved by noticing that the areas underneath the three peaks are in the same ratio 3 : 2 : 1 as the numbers

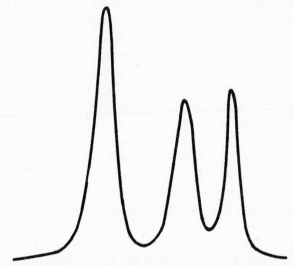

Fig. 39. Oscillograph trace of the proton resonance absorption signal from ethyl alcohol (Arnold, Dharmatti and Packard, 1951). Reading from left to right the three peaks correspond to resonance absorption in CH_3, CH_2 and OH groups. The separation between the outer peaks is 37 milligauss.

of protons in the groups. The shift is always relatively small for the proton, and the outer peaks are separated in this example by only 0·037 gauss in an applied field of 7600 gauss. This fine structure, characteristic of the compound, provides an important new tool in chemical analysis and identification (see for example Jarrett, Sadler and Shoolery, 1953).

5.7. Fine structure

We have just encountered one important type of fine structure in the nuclear magnetic resonance spectrum of liquid specimens, namely, that arising from different chemical shifts for chemically non-equivalent atoms, and exemplified by the three types of proton in ethyl alcohol.

Another type of fine structure was found by Proctor and Yu (1950 *b*, 1951), and later verified by Dharmatti and Weaver (1952 *c*), for the ^{121}Sb and ^{123}Sb resonances in an aqueous solution of antimony hexafluoantimonate ($NaSbF_6$) and hydrofluoric acid. The structure consisted of an equally spaced symmetrical set of seven lines with a spacing of about 1·8 gauss. For a liquid specimen the spectrum was thus most remarkably broad. Andrew (1951b) suggested that the octahedral $[SbF_6]^-$ ions may possess a high degree of rotational mobility about one axis on account of hydrogen bonding with the hydrogen fluoride molecules. The local magnetic field due to the six fluorine nuclei would then not average to zero (see §6·4) and would lead to a broad line as in crystals. This theory had the merit of giving a value for the splitting in good agreement with the experimental value. It now seems likely, however, that this quantitative agreement is coincidental, for similar fine structure has since been found by Gutowsky and McCall (1951) and Gutowsky, McCall and Slichter (1951, 1953) in a number of other liquids for which the rotational hindrance argument does not seem to be applicable. Such fine structure was also inferred by Hahn and Maxwell (1951, 1952) and McNeil, Slichter and Gutowsky (1951), who found that the small differences in frequency of the component lines gave rise to beats in the amplitude of the spin echo (see §5.5.2).

This type of fine structure is distinguished from that due to different chemical shifts† in that (a) it may occur even when there is only one resonant nucleus in each molecule (e.g. the ^{31}P resonance in $POClF_2$ gives a triple line), and (b) the splitting is independent of the strength H_0 of the steady field (Quinn and Brown, 1953), whereas the chemical shift splitting is proportional to H_0. Two other properties of this multiplet structure are:

(a) The energy-level displacements to which the splittings correspond are proportional to the product of the magnetic moments of the resonant nuclei and their non-identical neighbours.

(b) The number of lines and their relative intensities are determined by the number of possible spin states of the neighbours and their statistical weights. Thus for $POClF_2$, the ^{19}F resonance consists of two equally intense lines, since ^{31}P has spin number $\frac{1}{2}$, while the ^{31}P resonance consists of three lines, the central line

---

† Both types of fine structure may be observable simultaneously in certain cases, leading to a complicated resonance pattern (e.g. Shoolery, 1953).

having double the intensity of the two outer lines, corresponding to the four equally probable configurations of the two ^{19}F spins each of spin number $\frac{1}{2}$.

Gutowsky, McCall and Slichter (1951) and Hahn and Maxwell (1951) were led by these experimental facts to make independently the empirical postulate of an interaction between each resonant nucleus and its nuclear magnetic neighbours, proportional to the product of the magnetic moment vectors of the resonant nucleus and the neighbours. This interaction cannot be a direct interaction of the two magnetic moments, since, as is shown in §6.4, this averages to zero for a liquid specimen if the postulate of rotational hindrance in the liquid is abandoned. Instead, an indirect interaction was suggested by Gutowsky, McCall and Slichter (1951) via the orbital magnetic moments of the molecular electrons. It has been pointed out, however, by Ramsey and Purcell (1952) that this effect is too small to account for the observed splittings. They propose, alternatively, that nuclear moments interact via the exchange-coupled spin magnetic moments of the molecular electrons, a mechanism which can give splittings ten or a hundred times larger and comparable with those observed. In particular, they predicted the splitting to be expected for the HD molecule, for which the deuteron resonance should be a doublet and the proton resonance a triple line. This splitting has since been measured by Carr and Purcell (1952) and Wimett (1953a), and rather less accurately by Smaller, Yasaitis, Avery and Hutchison (1952); it has the small value of 43 c./s., corresponding to 10 milligauss for the proton resonance. A more complete theoretical analysis by Ramsey (1953) yields a value in good agreement with the observed splitting.†

We see that as in other branches of spectroscopy, improvement of resolving power has brought to light valuable fine details, and it is likely that attention will be paid to improving resolution still further. With an inhomogeneity of 1 milligauss over a specimen, which can be achieved with care, a resolution of 4 c./s. is obtained for protons. Bloch (1954) has pointed out that if the molecules within the specimen can be mechanically circulated over a range of

---

† This interaction is also present in solid specimens, but its effect on the spectrum is usually masked by the much larger magnetic dipolar interaction (§6.2). Nevertheless, this interaction may be responsible for small irregularities observed in the angular behaviour of the spectra of some solids (Pake, 1953).

inhomogeneity with a frequency greater than the inhomogeneity line width, then the resonance frequency for each molecule is determined by the average field each molecule experiences, and in consequence the resonance line is narrowed. Anderson and Arnold (1954) rotated a specimen of water at 25 revolutions per second, and found that the line width was reduced from 1·7 to about 0·1 milligauss.

Another type of fine structure is sometimes encountered when registering signals through a phase-sensitive amplifier with a recording meter. This structure is an unwanted distortion found when the frequency of modulation exceeds the resonance line width. It sets a practical limit to the sharpness of resonance lines when using very weak steady magnetic fields (Brown and Purcell, 1949; Brown, 1950), and has been discussed by Pound and Knight (1950), Smaller (1951) and Burgess and Brown (1952).

5.8. Water content in biological materials

It is frequently desirable for the biologist to know the water content of hygroscopic materials of biological origin, such as proteins, carbohydrates and vegetable tissue. One might expect that provided the environment of the water protons is independent of the water content, the size of the proton magnetic resonance signal in such materials should be proportional to the water content. For apple and potato tissue Shaw and Elsken (1950) have found a linear relationship between signal size and water content when the proportion of water in the material lay between about 10 and 80%. Nuclear magnetic resonance can therefore provide a basis for rapid determination of water content in such cases. The resonance spectrum consists of a narrow line due to the water superimposed on a broad line due to protons in the supporting solid (Shaw and Elsken, 1953; Shaw, Elsken and Kunsman, 1953).

In a somewhat similar experiment Jacobsohn, Anderson and Arnold (1954) investigated dilute solutions of deoxyribonucleic acid, in which it is supposed that large hydration shells are formed round the macromolecules. The protons in such shells lack the high mobility which endows the resonance of pure water with a very narrow line width, and instead give a line which is broad and therefore weak. By observing the strength of the strong narrow line due to the more mobile water protons outside the hydration shells, the proportion of water taken up in the shells was estimated.

5.9. Chemical reactions

The progress of chemical reactions in solution may be followed if some characteristic of the nuclear magnetic resonance changes during the course of the reaction. In cases where one of the reactants or of the products of reaction shows characteristic fine structure, observation of the strength of resonance associated with a particular molecular grouping can afford a means of observing the progress of the reaction.

In other cases the relaxation time T_1 may change as the reaction proceeds. Thus, for example, Hickmott and Selwood (1952) were able to follow the reduction of $EuCl_3$ to $EuCl_2$ in dilute hydrochloric acid solution by taking measurements of T_1 as the reduction proceeded. The effective magnetic moment μ_{ion} of the Eu^{3+} ion is some 15 times smaller than that of the Eu^{2+} ion. Since T_1 is inversely proportional to μ_{ion}^2 and to the ionic concentration (equation (5.37)), the relaxing effect of the Eu^{3+} ions is relatively negligible once the reaction has begun, and $(1/T_1)$ is then proportional to the Eu^{2+} concentration. In a similar manner Newman and Ogg (1951) were able to study the equilibrium between paramagnetic and diamagnetic species in dilute solutions of sodium in liquid ammonia.

Relaxation time studies can also furnish information in the study of solid paramagnetic catalysts for reactions in the liquid state. The catalytic activity depends upon the accessibility of the paramagnetic ions to molecules of the liquid. This accessibility in encounters between water molecules and the surfaces of the solid catalyst also determines the relaxing effect of the paramagnetic catalyst. A study of the variation of T_1 with catalyst concentration therefore provides information concerning the structure and mechanism of the catalyst (Spooner and Selwood, 1949; Selwood and Schroyer, 1950; Mooi and Selwood, 1952).

5.10. Nuclear magnetic resonance in gases

The first observation of nuclear magnetic resonance in a gaseous specimen was by Purcell, Pound and Bloembergen (1946) using hydrogen. The resonance line was sharp, as for a liquid, and of weak intensity on account of the low density of protons. Although the resonance has been observed with hydrogen at as low a pressure as 0·3 atmosphere (Packard and Weaver, 1952), it is usually necessary with gases to work with a pressure of at least 10 atmospheres to obtain adequate signal strength. At a pressure of 10 atmospheres the spin-lattice relaxation time for hydrogen is quite short, of the order 10^{-2} sec. As in the case of liquids relaxation is effected by the fluctuating local magnetic field. The local field at a given proton arises from two causes: that due to rotation of the molecule and that due to the other proton of the molecule. The corresponding interaction energies are fairly large and have been accurately measured in molecular beam experiments. These interactions involve the rotational quantum number of the molecule and its spatial quantization, which in a gas may be expected to change frequently as a result of molecular collisions. If the spatial quantization changes with each collision the correlation time τ_c for the local field will be of the order of the mean time between collisions, which in hydrogen at room temperature and at 10 atmospheres is about 10^{-11} sec. This situation is rather similar to that in a liquid of low viscosity, since $2\pi\nu_0\tau_c \ll 1$. We therefore expect to find, as with such liquids (see §§5.2–4), (a) that T_1 is inversely proportional to τ_c and therefore directly proportional to the pressure, and (b) that T_1 and T_2 are approximately equal. Purcell, Pound and Bloembergen (1946) confirmed that T_1 was proportional to the pressure between 10 and 30 atmospheres, while Packard and Weaver (1952) found from line width measurements that T_2 was proportional to pressure between 0·3 and 40 atmospheres. The latter workers also showed at the higher pressures that T_1 and T_2 were in fact about equal. Similar confirming evidence has been obtained for ethane gas by Verbrugge and Henry (1951).

The fluctuating intramolecular local magnetic fields which are thus responsible for spin-lattice relaxation for polyatomic molecules are absent for monatomic gases. In this case each nucleus only experiences an appreciable local magnetic field during a collision or near-collision, and in consequence the relaxation time for nuclei with spin number $\frac{1}{2}$ is very long. Brun, Oeser, Staub and Telschow (1954 b) find a relaxation time of the order of 10^3 sec. for ^{129}Xe in xenon gas at a pressure of 50 atmospheres. Relaxation was more rapid for ^{131}Xe (spin number $\frac{3}{2}$), which possesses a quadru-pole moment (see §8.2). The long relaxation time for monatomic gases containing nuclei with spin number $\frac{1}{2}$ can be reduced by addition of a catalyst. H. L. Anderson (1949) measured the nuclear magnetic moment of ^{3}He using a specimen containing equal volumes of ^{3}He and oxygen compressed to 20 atmospheres. The oxygen molecule has a permanent magnetic moment about 10^3 times greater than that of a ^{3}He nucleus, causing T_1 to be cut down to about 1 sec. In order to reduce T_1 for ^{129}Xe, Proctor and Yu (1951) adopted an alternative catalyst, suggested by Bloch (1951b). The specimen container is filled with finely powdered ferric oxide and with xenon gas at a pressure of 12 atmospheres. There exist strong magnetic fields close to the surfaces of the paramagnetic powder. Consequently the collision of a gas atom with a surface can cause a similar reorientation of its nuclear moment to that achieved in an impact with a molecule of admixed oxygen. For the powder used the shortening of T_1 was the same as would be obtained with a partial pressure of about 30 atmospheres of oxygen gas (Bloch, 1951b).

Nuclear magnetic resonance studies have also been made of hydrogen gas occluded in metals. With the alkali and alkaline earth metals hydrogen forms salt-like hydrides in stoichiometric pro-portions. The proton resonance line width for these compounds is broad (Garstens, 1950), in the manner characteristic of rigid solids (Chapter 6). With the transition metals the situation is rather different; the hydrogen may be absorbed into interstitial positions in the metallic lattice to a degree which depends upon the gas pressure and the temperature. With titanium, Garstens (1950) finds a broad line as with the salt-like hydrides of the alkali and alkaline earth metals, which suggests that the hydrogen atoms take up fixed interstitial positions. On the other hand, with tantalum the line is quite narrow suggesting that the hydrogen atoms are free to migrate through the lattice from one site to another. At low temp-eratures this motion is effectively frozen however (Garstens, 1951). A more complete study of hydrogen occluded in palladium has been made by Norberg (1952), who similarly attributes the narrow line width obtained to the well-known migration of the protons through this metal.

5.11. Fermi-Dirac degeneracy in ³He

An experiment of considerable interest was carried out on ³He in the gaseous and liquid states between 4·2 and 0·23° K. by Fairbank, Ard, Dehmelt, Gordy and Williams (1953) and Fairbank, Ard and Walters (1954). If liquid ³He behaves as an ideal Fermi–Dirac gas, the molar magnetic susceptibility should depart from the inverse temperature proportionality indicated by equation (2.33). It will be recalled that the derivation of (2.33) assumed the operation of classical statistics, whereas for ³He atoms, whose nuclear spin number is ½, one might perhaps expect Fermi–Dirac statistics to apply. The difference of behaviour predicted by the two kinds of statistics is only important near or below the degeneracy temperature, which, for particles of this mass and density, is about 5° K.

As equation (2.40) shows, the nuclear magnetic resonance signal strength is a measure of the volume nuclear magnetic susceptibility provided saturation is avoided, and provided the resonance line width and line shape remain unaltered. Taken with the known temperature variation of the density of ³He, the signal strength thus gives a measure of the molar nuclear susceptibility. Signal strength measurements indicated that down to about 1° K. the molar susceptibility was inversely proportional to the temperature, but that below 1° K. there was a pronounced departure from this behaviour. The relative values of molar susceptibility which were found, fit the theoretical curve for a Fermi–Dirac gas provided that the degeneracy temperature is empirically adjusted to be 0·45° K., some ten times lower than the value expected if ³He were an ideal Fermi–Dirac gas.

Notes added in proof

(1) In a recent paper Kubo and Tomita (1954) have re-examined and revised some of the equations developed by Bloembergen, Purcell and Pound (1948) which are quoted in §§5.2 and 5.3. The general conclusions are not changed however.

(2) Perhaps the most notable advance in nuclear magnetic resonance since the manuscript of this book was completed has been the examination and use of the fine structure of the resonance lines for liquids (§5.7). For examples of this high resolution work the reader is referred to papers read at the recent conference on Microwave and Radiofrequency Spectroscopy held in Cambridge, April 1955, and published in *Discussions of the Faraday Society* (1955).

CHAPTER 6

NUCLEAR MAGNETIC RESONANCE IN NON-METALLIC SOLIDS

6.1. Introduction

For our purposes the essential difference between liquids and gases on the one hand and solids on the other, is that in the former case the molecules possess great translational and rotational freedom, whereas in the solid state they generally possess very little. This lack of molecular motion has two important consequences. In the first place the local magnetic field due to neighbouring magnetic dipoles is static, and as we have already briefly mentioned in §2.3, causes the resonance absorption line width to be broad. On the other hand, as we saw in the previous chapter, in liquids and gases the rapid motion of the molecules smooths out the local field to a very small average value so that the line is quite narrow. This time-varying local field in liquids is also the all-important source of spin-lattice relaxation. Since this is usually absent in solids a second important consequence is that solids generally have much longer spin-lattice relaxation times than liquids. It follows therefore that for solids the spin-lattice relaxation time T_1 is usually much longer than the spin-spin interaction time T_2, whereas for liquids the two times are frequently of the same order (§5.3). Thus, for example, it is quite possible for a crystal to have $T_1 \sim 10^4$ and $T_2 \sim 10^{-5}$ sec. There are, however, exceptions to these generalizations, and we shall later encounter solids within which the molecules are endowed with a remarkable degree of freedom, and for which the resonance line width is as narrow as that of a liquid and the spin-lattice relaxation time quite short.

In this chapter we shall deal only with non-metallic solids, postponing the rather special case of metals to Chapter 7. Quadrupole effects in solids are deferred until Chapter 8; this means that we shall be largely concerned in this chapter with nuclei of spin number $\frac{1}{2}$ only. We shall first turn our attention to solids whose crystal structure is virtually rigid, and will examine the resonance

L

line shape in some detail. We shall discover that the line shape is an important tool in the analysis of crystal and molecular structures. We shall then consider the effect on the resonance line of molecular reorientation and diffusion in solids in which these processes occur. These processes will be found to be responsible for spin-lattice relaxation as they were in liquids. We shall then inquire into the relaxation mechanism for solids in which these processes are absent, and finally close the chapter with consideration of a number of special topics. It is not possible to discuss all the work which has been done on solids; however, a list of solids which have been investigated is given, with references, in Appendix 4.

6.2. Nuclear magnetic resonance absorption spectrum for rigid structures

6.2.1. Spectrum for systems of two identical nuclei. We open our discussion of line shapes for solid specimens with a consideration of solids in which nuclear magnets occur in pairs, a situation elegantly studied by Pake (1948). Examples of such solids are many hydrated salts, such as gypsum, $CaSO_4 . 2H_2O$. Apart from certain isotopes of calcium, sulphur and oxygen of meagre abundance, the only magnetic dipoles present in the crystal lattice of gypsum are the protons of the water molecules. The experimental evidence suggests that the arrangement of atoms in the crystal lattice is rigid at room temperature, save, of course, for vibration of the atoms about their equilibrium positions. We have therefore, for magnetic purposes, an assembly of identical magnetic dipoles arranged in pairs at fixed sites in the crystal. The separation of the protons within each pair is about $1 \cdot 58$ Å., while next nearest neighbouring protons are about $2 \cdot 8$ Å. away. Since the interaction between magnetic dipoles falls off as the inverse cube of the separating distance, each proton finds itself predominantly in the local field of its partner in the same water molecule. From a classical viewpoint the component of this local field parallel to the applied steady field H_0 is $\pm \mu r^{-3}(3 \cos^2 \theta - 1)$, where μ is the proton moment, r is the pair separation and θ is the angle between H_0 and r; the positive or negative sign is to be taken according as the partner proton is respectively parallel or antiparallel to H_0. If, therefore, to a first approximation, we neglect

the local field produced by the more remote neighbours, we see that each proton is situated in a resultant field of strength H given by

$$H = H_0 \pm \mu r^{-3}(3\cos^2\theta - 1). \qquad (6.1)$$

From a given proton pair and the corresponding pairs in all other unit cells in a single crystal of gypsum, we therefore expect two proton resonance lines equally spaced about the resonance field strength $H^* = h\nu/2\mu$ by an amount $\mu r^{-3}(3\cos^2\theta - 1)$, where ν is the radiofrequency. The unit cell of gypsum contains two types of proton pair with, in general, different values of θ. A single crystal should therefore give two pairs of symmetrically disposed resonance lines. Finally, we remember the smaller local field due to all other neighbours; this has the effect of broadening the four lines. Fig. 40 shows measured spectra of gypsum obtained by Pake

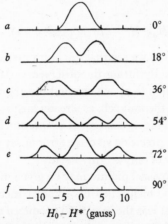

Fig. 40. Resonance absorption spectra for various directions of the applied field H_0 in the (001) plane of a gypsum single crystal (Pake, 1948). The angle between H_0 and the [100] direction is stated with each spectrum.

(1948) for various orientations of a single crystal about its (001) direction. At some settings the two pairs of lines are clearly evident, as, for example, at setting (d); at other settings, for example, (b) and (f), the separations for the two types of proton pair are about the same so that only two peaks are seen; at setting (e) one pair has coalesced at the centre. By examining the variation with orientation of the displacement of the peaks from the centre, and using

(6.1) the angular disposition of the proton pairs in the unit cell is found.

It is to be noticed that the spectra are all quite broad. It would appear from (6.1) that the maximum displacements of a peak from the centre of the pattern should be $2\mu r^{-3}$, and this expression certainly leads one to expect a pattern some ten gauss broad as observed. We might use this expression to find r, but before doing so it is necessary to recall that (6.1) is based upon a classical evaluation of the local field, and we must therefore first find the quantum-mechanical equivalent of (6.1). Pake (1948) treated the magnetic dipole-dipole interaction between the two protons in a pair as a perturbation of the energy levels of the protons in the applied field H_0. By a straightforward first-order perturbation calculation, which is carried out in Appendix 6, he found the new energy levels, and from these found that the resonance peak should occur at field strengths given by

$$H_0 = H^* \pm \tfrac{3}{2}\mu r^{-3}(3\cos^2\theta - 1). \tag{6.2}$$

Comparing with (6.1) we see that quantum mechanics supplies an additional factor $\tfrac{3}{2}$. By fitting the observed variation of peak displacement to equation (6.2) one can find the angular disposition of the proton pairs in the unit cell, as already mentioned, and also the pair separation r. This measurement of internuclear distance may be made with some precision since an error of say 6% in measuring μr^{-3} causes an error of only 2% in r itself.

The following physical picture may be given of the origin of the factor $\tfrac{3}{2}$. In §2.3 we mentioned two dipolar mechanisms for line broadening: the first was the local field produced at one nucleus by the z component of the other; the second arose from the limitation of the lifetime of the spin state by simultaneous spin-exchange transitions on the part of neighbouring identical nuclei mutually induced by their precession. This second process is responsible for the factor $\tfrac{3}{2}$. If the two nuclei in each pair are not identical their precession frequencies differ, and the spin-exchange process does not occur; in this case Pake's analysis (see Appendix 6) shows that the factor $\tfrac{3}{2}$ is absent, and the semi-classical expression (6.1) is correct.

Our specimen containing identical nuclear dipole pairs may not be in monocrystalline form. If, instead, it is polycrystalline it can often be assumed that there are a great many crystal grains whose individual crystal axes are distributed randomly over all directions. The spectrum for this specimen is therefore the sum of the spectra for the individual grains, and we therefore expect the fine structure to be smeared out. Since the orientation of the dipole pairs is isotropically distributed, the fraction of pairs for which θ lies in the interval $d\theta$ is $d(\cos\theta)$. Let us write the normalized line-shape function, which describes the absorption signal as a function of field strength, as $g(h)$, where h is defined by

$$h = H_0 - H^*. \tag{6.3}$$

Then the contribution of each of the two component lines for a single crystal to the spectrum for polycrystalline material is

$$g(h) = \frac{1}{2}\frac{d(\cos\theta)}{dh}, \tag{6.4}$$

the factor $\frac{1}{2}$ entering since each of the two lines is equally probable. Now from (6.2) and (6.3) we have

$$h = \pm\tfrac{3}{2}\mu r^{-3}(3\cos^2\theta - 1). \tag{6.5}$$

Expressing $\cos\theta$ as a function of h from (6.5), and substituting in (6.4) we get

$$g(h) = (6\sqrt{3}\,\mu r^{-3})^{-1}\,[1 \pm h/(\tfrac{3}{2}\mu r^{-3})]^{-\frac{1}{2}}. \tag{6.6}$$

The $+$ sign is to be taken for $-\tfrac{3}{2}\mu r^{-3} < h < 3\mu r^{-3}$, and the $-$ sign for $-3\mu r^{-3} < h < \tfrac{3}{2}\mu r^{-3}$. The line shape obtained from (6.6) is plotted as the broken curve in fig. 41, and is seen to retain a characteristic structure. This line shape so far neglects the broadening of the line by the local field of all other neighbours. After this has been taken into account for gypsum the full line in fig. 41 is obtained, and some evidence of the original fine structure is still to be discerned from the double-humped curve. The experimental curve for polycrystalline gypsum is in good agreement with this curve. From the shape of the line and the separation of the humps, the interproton distance in the water molecules can be found, though the accuracy is less than is afforded by use of a single crystal. The use of polycrystalline material precludes the measurement of the angular disposition of the proton pairs.

A number of hydrated salts have been examined, either in mono- or polycrystalline form (see Appendix 4). The characteristic double-peaked curve is also found for other materials containing close

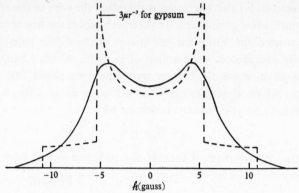

Fig. 41. The broken line shows the calculated resonance line shape for the protons in polycrystalline gypsum $CaSO_4 . 2H_2O$, taking into account nearest neighbour interactions only. The full line is obtained after taking into account the interaction of other neighbours.

pairs, as, for example, for 1.2-dichloroethane (Gutowsky, Kistia-kowsky, Pake and Purcell, 1949), from the spectrum of which the interproton distance in the CH_2Cl group was found.

6.2.2. Spectrum for systems of three identical nuclei. Some materials contain nuclei clustered in relatively separate groups of three. As examples, a number of solid organic compounds contain methyl groups (CH_3), and a number of acid hydrates contain oxonium ions (H_3O^+), each containing three protons at the vertices of an equilateral triangle. The more complicated perturbation analysis for the interaction of three nuclei has been given by Andrew and Bersohn (1950) for identical nuclei of spin $\frac{1}{2}$ at the corners of a triangle of general shape. Suppose we have a single crystal containing identical triangular groups similarly oriented. Neglecting at first the broadening caused by nuclei in other tri-angular groups, the analysis shows that the spectrum consists of a central line (at $h=0$), and three pairs of lines symmetrically dis-posed about the centre. The separation of these lines and their relative intensity are functions of the size and shape of the triangles and their orientation relative to the applied field H_0.

We shall not consider further the behaviour for a single crystal, since all experimental work has been concerned with polycrystalline material. The spectrum for polycrystalline material is obtained by summing the spectra for isotropically oriented crystal grains, as was done in the preceding section for pairs of nuclei. If the triangles are equilateral the line has the complicated shape shown in fig. 42 a. The rich detail of this spectrum is largely removed by the

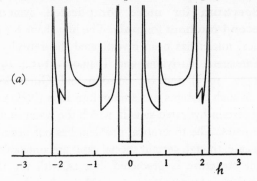

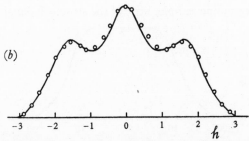

Fig. 42. Theoretical resonance line shapes for an assembly of randomly oriented groups of three identical nuclei of spin number $\frac{1}{2}$ situated at the corners of equilateral triangles of side r_0, (a) taking into account only magnetic interactions within each group, (b) taking into account interactions between nuclei in different groups also. The units of h are $\frac{3}{2}\mu r_0^{-3}$, where μ is the nuclear magnetic moment. Also shown in (b) are experimental points for the proton resonance from the oxonium ions (H_3O^+) in nitric acid monohydrate at 90°K. (Richards and Smith, 1951), for which the unit of h is 4·15 gauss.

broadening caused by nuclear dipoles in neighbouring groups; the resulting spectrum for the proton resonance in nitric acid monohydrate, which contains oxonium ions, is shown in fig. 42 b (Richards and Smith, 1951). By fitting experimental spectra to theoretical line shapes of this form the arrangement of nuclei in the oxonium ion has been deduced by Richards and Smith (1951) and by Kakiuchi, Shono, Komatsu and Kigoshi (1951, 1952).

The case of a linear three-spin system consisting of two identical nuclei with a different nucleus between them has been analysed by Waugh, Humphrey and Yost (1953), and the results applied to the bifluoride ion in KHF_2 and $NaHF_2$.

6.2.3. Spectrum for more complicated systems: Van Vleck's second-moment formula. The line shape for groups of four identical nuclei has been investigated theoretically by Itoh, Kusaka, Yamagata, Kiriyama and Ibamoto (1952, 1953 *b*), by Tomita (1952, 1953) and by Bersohn and Gutowsky (1954). Examples of such systems are ammonium ions (NH_4^+), methane (CH_4), and certain hydrated salts in which the water molecules lie together in pairs. The theoretical problem has not been solved in the general case, though certain special cases are amenable. A complicated line structure is predicted, but as with the three-spin system, broadening by next nearest neighbours largely removes the detail, and when averaged over all orientations to give the line shape for a polycrystalline sample, little of the structure remains. Fig. 43

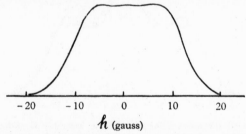

h (gauss)

Fig. 43. Resonance line shape for protons in polycrystalline ammonium chloride at 95°K. (Gutowsky, Kistiakowsky, Pake and Purcell, 1949).

shows the proton magnetic resonance spectrum for polycrystalline ammonium chloride.

It seems clear that for more general systems in which the nuclei are not conveniently localized in small groups the task of calculating the line shape would be formidable, and, moreover, the result would be less rewarding since the line structure becomes so complex as to be unresolved in detail. Nevertheless, although the line shape for general systems cannot be calculated, Van Vleck (1948) has shown that the moments of the spectrum can be readily calculated.† If the normalized line shape is described by the function $g(h)$ as before,‡ the nth moment, which we will call S_n, is defined as

$$S_n = \int_{-\infty}^{\infty} h^n g(h)\, dh. \qquad (6.7)$$

Since $g(h)$ is an even function for magnetic dipolar broadening the odd-numbered moments are all zero. Van Vleck calculated the second and fourth moments in the general case by a diagonal sum method. The fourth moment is a rather complicated expression, and is not much used. For a crystal containing only one species of magnetic nucleus his result for the second moment, S, which is just the mean-square width of the spectrum about its centre, is

$$S = \tfrac{3}{2} I(I+1) g^2 \mu_0^2 \mathcal{N}^{-1} \sum_{j>k} U_{jk}^2. \qquad (6.8)$$

As before, I is the spin number of the nuclei, μ_0 is the nuclear magneton, $g\mu_0 I$ is the nuclear magnetic moment, $\mathcal{N}$ is the number of magnetic nuclei in the system over which the sum is taken, and for a single crystal

$$U_{jk} = (3 \cos^2 \theta_{jk} - 1) r_{jk}^{-3}, \qquad (6.9)$$

where r_{jk} is the length of the vector joining nuclei j and k, and θ_{jk} is the angle between this vector and the applied magnetic field $\mathbf{H}_0$. Thus if the co-ordinates in the crystal lattice of all the magnetic dipoles are known, U_{jk} for every pair of nuclei in the crystal may be calculated from (6.9), and using (6.8) the second moment is obtained.

Strictly the number of nuclei $\mathcal{N}$ is the total number of nuclei in the crystal, and the sum in (6.8) is taken over all $\tfrac{1}{2}\mathcal{N}(\mathcal{N}-1)$ nuclear pairs in the crystal. However, since the inverse sixth power of distance enters into each U_{jk}^2 (see (6.9)), important contributions to the sum are only obtained from pairs of nuclei in the same unit cell or in adjacent cells. The environment of all unit cells is therefore effectively identical; the number $\mathcal{N}$ may therefore be reduced to the number of nuclei in one unit cell, and the sum taken over all the nuclei in this cell and their neighbours, both inside and outside

---

† Second-moment formulae have also been derived by Pryce and Stevens (1950) and by Kambe and Usui (1952). The relationship between the moments of a spectrum and its Fourier transform has been discussed by Yokota (1952).

‡ This is the same shape function as that defined in §2.3, except that here the field strength is taken as the variable instead of the frequency.

the cell. Frequently the symmetry of the unit cell provides two or more nuclei with the same environment; in this case $\mathcal{N}$ may be reduced still further. Moreover, it is not necessary to calculate r_{jk} and θ_{jk} for the more remote neighbours; it usually suffices to treat exactly the neighbours up to about 5 Å. distant, and to convert the sum for the remainder into an integral, assuming the more distant neighbours to be uniformly distributed with known density.

If the specimen is polycrystalline, the absorption spectrum is the sum for the assembly of isotropically oriented crystal grains, and its second moment is therefore the average of the second moments of the individual grains. Since the isotropic average of $(3 \cos^2 \theta_{jk} - 1)^2$ is readily shown to be $\frac{4}{5}$, we find using (6.8) and (6.9) that the second moment is

$$S = \tfrac{6}{5} I(I+1) g^2 \mu_0^2 \mathcal{N}^{-1} \sum_{j>k} r_{jk}^{-6}. \tag{6.10}$$

If the crystal lattice contains other species of magnetic nuclei besides those at resonance, Van Vleck's second-moment formula contains the following additional term, which must be added to (6.8):

$$\tfrac{1}{3} \mu_0^2 \mathcal{N}^{-1} \sum_{j,f} I_f(I_f+1) g_f^2 U_{jf}^2, \tag{6.11}$$

where I_f is the spin number of nucleus f of another species, $g_f \mu_0 I_f$ is its magnetic moment and for a single crystal

$$U_{jf} = (3 \cos^2 \theta_{jf} - 1) r_{jf}^{-3}, \tag{6.12}$$

where r_{jf} is the length of the vector joining nuclei j and f and θ_{jf} is the angle between this vector and the applied magnetic field $\mathbf{H}_0$. For a polycrystalline specimen U_{jf}^2 must be replaced by its mean value in (6.11), and the second-moment formula therefore becomes

$$S = \tfrac{6}{5} I(I+1) g^2 \mu_0^2 \mathcal{N}^{-1} \sum_{j>k} r_{jk}^{-6} + \tfrac{4}{15} \mu_0^2 \mathcal{N}^{-1} \sum_{j,f} I_f(I_f+1) g_f^2 r_{jf}^{-6}. \tag{6.13}$$

It is to be noticed that the numerical coefficient in (6.8) for the contributions to the second moment from pairs of identical nuclei is $\frac{9}{2}$ times greater than that in (6.11) for unlike pairs. One expects a trivial difference of a factor 2 in these coefficients since the sum in (6.8) is arranged, by the requirement $j>k$, to count each pair of nuclei once only, so that the real difference in coefficients is a factor $\frac{9}{4}$. This is merely the square of the factor $\frac{3}{2}$ we encountered in

§6.2.1, which may be said to arise from the mutual spin exchange process which can take place between identical nuclei only.

The second-moment formula (6.8) and the fourth-moment formula which we have not quoted have been verified experimentally by Pake and Purcell (1948) using a single crystal of calcium fluoride, CaF_2. Support for the different numerical coefficient in (6.11) and (6.13) for unlike nuclear neighbours is provided by the work of Pake (1948, see also Van Vleck, 1948) on the ^{19}F resonance in KHF_2 and $KF.2H_2O$. As we shall see in the next section Van Vleck's second-moment formula has proved to be an important tool in structure determination, and the consistent results which have been obtained with its use in a large number of investigations testify to its correctness.

6.3. Structural determinations from the nuclear magnetic resonance absorption spectrum

We have seen in the previous section that if a nuclear magnetic resonance spectrum owes its width to nuclear magnetic dipolar broadening, the second moment or mean-square width of the spectrum may be calculated from a knowledge of the disposition of all nuclei of non-zero spin in the crystal lattice, and of their spin numbers and magnetic moments. Since the spin numbers and magnetic moments are usually known, and since the second moment may be derived from the experimentally determined spectrum, it follows that information concerning the disposition of the nuclei may be obtained. The information is not usually sufficient to allow a complete determination of a crystal structure to be made, but very often suffices to fill in the details of the picture provided by other methods. The method is especially valuable in locating the positions of hydrogen nuclei in crystals since hydrogen atoms are very weak scatterers of X-rays, and cannot readily be located by the X-ray method.

The procedure is perhaps best understood by means of a specific example. The spectrum for polycrystalline ammonium chloride at 88° K. is shown in fig. 43, and its second moment is 53·6† gauss²

† After allowing for the effect of zero-point torsional oscillation of the ammonium ions, which, as is mentioned at the end of this section, causes the second moment to be less than it would be if the ions were quite stationary.

(Gutowsky, Pake and Bersohn, 1954). The crystal structure is cubic with known cell size; the positions of the nitrogen and chlorine nuclei are therefore known. The four equal N—H bonds in the ammonium ion may be assumed to be tetrahedrally distributed, but their length is not known. Using (6.13) this distance is found to be 1·038 Å.[†] Since the magnetic moments of ^{14}N, ^{35}Cl and ^{37}Cl are small compared with the proton moment, and since the square of the magnetic moment enters the second-moment formulae, the second moment is mainly determined by the proton-proton interactions. In fact of the total of 53·6 gauss^2 only 1·8 gauss^2 is contributed by the second term in (6.13). Of the remaining 51·8 gauss^2, only 6·5 gauss^2 is contributed by pairs of protons in different ammonium ions, a dominant contribution of 45·3 gauss^2 coming from the more closely spaced intra-ionic proton pairs. The accuracy of the measured second moment was about 2%; on account of the inverse sixth power dependence on distance, the N—H bond is thus determined with an accuracy of about 0·4% or ±0·004 Å.

It is to be noted that if one works with polycrystalline material, there is only one measured quantity, namely, the second moment. Consequently only one parameter of the crystal structure can be determined unambiguously, and all others must be known from other sources. More information is usually to be obtained from single crystals, especially when the anisotropy of the second moment is marked. Moreover, the spectrum for a single crystal is more likely to display fine structure, and this, as we saw in §§6.2.1 and 6.2.2, can give valuable information about the presence of simple groups of nuclei in the structure. As an example, Andrew and Hyndman (1953) have been able to show from the anisotropy of the second moment, using a single crystal, that the urea molecule, $OC(NH_2)_2$, is wholly planar.

It sometimes happens that when investigating molecular compounds, one has a fairly good idea of the molecular structure, but has no information concerning the crystal structure. If one then postulates a certain arrangement of the hydrogen atoms in the

[†] Bersohn and Gutowsky (1954) have also examined ammonium chloride in monocrystalline form, and find the N—H bond length to be 1·032 ±0·005 Å. These values are in excellent agreement with the value 1·03 ±0·02 Å. obtained by the neutron diffraction method by Levy and Peterson (1952).

molecule one can calculate the intramolecular contribution to the second moment, using (6.8) and (6.11) or (6.13), but one is handicapped in comparing this with the experimental second moment by lack of knowledge of the intermolecular contribution. It is sometimes possible to estimate the intermolecular contribution by comparison with similar compounds whose crystal structure is known. This comparison may be put on a semi-quantitative basis by an argument which recalls that the expression for the dominant term in the lattice energy of apolar molecular solids, namely, the electric dipole-dipole term of the van der Waals energy of attraction, contains sums $\sum r_{jk}^{-6}$ of the same form as those entering in (6.13) (Andrew and Eades, 1953b).

The difficulty is better overcome by the method of isotopic substitution where it can be applied. The method was devised by Andrew and Eades (1953a, c), who applied it to polycrystalline benzene. Experimental proton second-moment values were obtained for ordinary benzene, C_6H_6, and for 1.3.5-trideutero-benzene, $C_6H_3D_3$. The substitution reduces the intramolecular and the intermolecular contributions by very different factors both of which are calculable. From the two measured second moments the intra- and intermolecular contributions can therefore be evaluated separately.

Certain experimental aspects of these structural investigations are worthy of note. By contrast with liquids the spectrum is broad (T_2 short) and the spin-lattice relaxation time T_1 is long. We see from equation (3.26) that both these features operate against the production of a good signal-to-noise ratio. Indeed, the spectrum can rarely be presented on an oscillograph, and instead the first derivative of the spectrum is usually recorded using a phase-sensitive amplifier in the manner described in §3.2. The second moment may be computed directly from this derivative since, from (6.7), we have

$$S = \int_{-\infty}^{\infty} h^2 g(h) dh = -\frac{1}{3} \int_{-\infty}^{\infty} h^3 \left(\frac{dg}{dh}\right) dh, \qquad (6.14)$$

by partial integration, remembering that $g(h)$ vanishes except when h is close to zero. A signal proportional to the absorption itself, rather than to its derivative, may be obtained by the use of a square-wave field modulation of large amplitude, instead of the sine-wave of small amplitude envisaged here; this alternative does not seem to have been much used however.

A number of precautions are essential if accurate values of second moment are to be obtained.

(a) The radiofrequency power level must be low enough to prevent distortion of the spectrum by saturation effects.

(b) The spectrum must not be unduly broadened by magnetic field inhomogeneity.

(c) When plotting the derivative of the line shape, the field modulation sweep must be small compared with the line width. However, the signal-to-noise ratio does not always allow this sweep to be made as small as one would like. A simple and exact correction formula for spurious broadening due to this cause has been given by Andrew (1953) (a more approximate correction had previously been given by Perlman and Bloom (1953)).

(d) The signal-to-noise ratio must be good, since the second moment can be greatly influenced by the wings of the absorption line where the signal is weak.

(e) The positions of the atoms in the crystal lattice must remain fixed. As we shall see in §§6.4 and 6.5, there is considerable molecular motion, rotational and sometimes translational, in many solids, particularly at temperatures close to their melting-points. Such motion has a profound effect upon the spectrum and its second moment. It is therefore essential to work at temperatures so low that such molecular motion either does not take place or takes place too slowly to affect the spectrum. It was for this reason that the spectrum of ammonium chloride, referred to earlier in this section, was taken at 78° K. However, even at the lowest temperatures there is zero-point motion which may be appreciable, necessitating a correction (Gutowsky, Pake and Bersohn, 1954).

Finally, it should be said that as a method for elucidating crystal structures containing light atoms, nuclear magnetic resonance is not so powerful a tool as its rival, the neutron diffraction method. Nevertheless, for structural problems which are not too ambitious it is equal to its task, and, moreover, the facilities required are conspicuously more modest and are available to all. Furthermore, as we shall see in the next sections, the nuclear magnetic resonance method comes into its own for the many crystals in which the atoms, molecules or groups are not rigid but instead have rotational or translational motion.

6.4. Hindered rotation in solids

In many solids constituent molecules or atomic groups undergo rotational motion about one or more axes with a frequency which increases with temperature. Such motion modifies the interactions between nuclear magnetic dipoles in the rotating groups and causes a considerable change in the resonance absorption spectrum. Looking at the situation in terms of the local field produced at any resonant nucleus by its neighbouring dipoles, we observe that this field is now time-varying, and that if the variation is sufficiently rapid, the time average of the local field must be taken to express the dipole interaction. Since the time average over all permitted orientations of the dipole pairs can in general be expected to be less than the steady local field for a rigid system, the spectrum may be expected to become more narrow when the rotation sets in. Indeed, this is a step towards the behaviour for liquids discussed in Chapter 5. There we found that the rapid random reorientation of the molecules together with their translational motion, caused the average value of the local field to be extremely small, with the result that the resonance line was very narrow and frequently determined in practice by the inhomogeneity of the steady magnetic field.

A narrowing of the absorption line with increasing temperature has been observed in many solids and has been ascribed to molecular motion within the crystal lattice. The application of nuclear magnetic resonance to processes of this kind was pioneered by Alpert (1947, 1949), and was put on a quantitative basis by Gutowsky and Pake (1950), who built their work upon the important theoretical analysis of Van Vleck (1948) of the dipolar broadening of resonance lines, which we have just discussed in §6.2.3.

Molecular rotation in crystals is sometimes sufficiently marked to bring about anomalies in the specific heat and transitions in the crystal structure. Such transitions can, however, have other causes, and unambiguous evidence of molecular rotation had hitherto been obtainable mainly from X-ray or dielectric measurements. X-ray methods only give such information if each molecule spends a large proportion of its time rotating, whereas the nuclear resonance line width is affected even though each molecule spends only a small fraction of its time rotating, as we shall see presently. Dielectric measurements do not suffer from this lack of sensitivity, but they are only applicable to polar molecules, while non-polar molecules are more likely to rotate on account of their greater symmetry, and their freedom from polar binding forces.

Before we proceed further we must define rather more closely the processes so far described rather loosely as 'rotation'. Let us consider, as an example, solid benzene. The benzene molecule possesses a six-fold (hexad) axis of symmetry normal to the plane of the molecule. There are in consequence six equivalent positions for each molecule in the crystal lattice, since a reorientation of any molecule through successive angles of 60° does not alter the structure. The molecules are, however, constrained to their equilibrium positions by intermolecular forces, and rotation of any molecule from one equilibrium position to another is restricted by a potential barrier. If the height of this hindering barrier is V per mole, a number of molecules proportional to $\exp(-V/RT)$ have sufficient energy at any instant to surmount the barrier, where R is the gas constant per mole. There will thus be a continual reorientation of all the molecules in the lattice about their hexad axes. In some solids the barrier restricting reorientation is so high that for practical purposes the structure may be considered rigid. On the other hand, in other solids the barrier may be so low that there is continuous, almost free, rotation. In certain solids, as for example, benzene, reorientation takes place almost exclusively about one axis; in others, for example salts containing the highly symmetrical ammonium ion, reorientation about a number of axes is equally probable. Reorientation may also take place by quantum-mechanical tunnelling through the hindering potential barrier, although this process does not seem to have been important for most of the solids studied.† Since all these processes can modify the resonance spectrum we shall use the word 'reorientation' to describe them, rather than the word 'rotation', which might be taken to refer only to the special case of free rotation.

† Tunnelling appears to be important in potassium dihydrogen phosphate and arsenate, however (Newman, 1950a, and see §6.7.3).

Even when the barrier is too high to be surmounted or penetrated the local magnetic field may be somewhat modified by rotational oscillation of the molecules or groups (Andrew, 1950).

If molecular reorientation takes place about one axis only, it is convenient to describe the random reorientation process by means of a correlation time τ_c, as we did for molecular reorientation in liquids in Chapter 5. Since the number of molecules with energy sufficient to surmount the potential barrier V is proportional to $\exp(-V/RT)$, we expect τ_c to vary with temperature according to the equation

$$\tau_c = \tau_0 \exp(V/RT), \qquad (6.15)$$

where τ_0 is a constant. It is frequently convenient to introduce a reorientation rate ν_c defined† by the equation

$$2\pi\nu_c\tau_c = 1, \qquad (6.16)$$

so that
$$\nu_c = \nu_0 \exp(-V/RT), \qquad (6.17)$$

where
$$2\pi\nu_0\tau_0 = 1. \qquad (6.18)$$

This reorientation rate thus increases with temperature, and we may therefore inquire what is the minimum reorientation rate which must be achieved before the spectrum is affected, or, in other words, before the nuclear magnetic interactions cease to be effectively static. This question has already been discussed in §5.3 when dealing with molecular reorientation in liquids, and it was taken that the resonance line narrows when the reorientation rate is of the order of the frequency line width itself; in practice this frequency is usually about 10^4 or 10^5 c./s.

6.4.1. Systems of two identical nuclei. Suppose we have relatively isolated systems of two nuclei of the type discussed in §6.2.1, which are reorienting about some axis at a rate ν_c much greater than the frequency line width. In fig. 44 OP is the vector $\mathbf{r}_{jk}$ joining the two nuclei j and k, and making an angle θ_{jk} with the applied field $\mathbf{H}_0$; ON is the axis of reorientation making an angle θ' with $\mathbf{H}_0$, and an angle γ_{jk} with $\mathbf{r}_{jk}$. If reorientation were not taking place, then neglecting the broadening effect of remoter nuclear

† It might be argued that a more appropriate definition of reorientation rate would be $1/\tau_c$. The definition adopted here is that introduced by Gutowsky and Pake (1950).

M

neighbours, the spectrum consists of two lines, as we saw in §6.2.1, given by (equations (6.2) and (6.3))

$$h = \pm \tfrac{3}{2} \mu r_{jk}^{-3} (3 \cos^2 \theta_{jk} - 1). \tag{6.19}$$

The reorientation causes θ_{jk} to vary with time as the vector OP takes up positions on the conical surface shown in fig. 44, and it

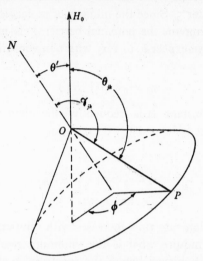

Fig. 44. Diagram illustrating the motion of an internuclear vector OP about an axis ON (Gutowsky and Pake, 1950).

becomes necessary to take the time-average value of the factor $(3 \cos^2 \theta_{jk} - 1)$ in (6.19).

If the reorientation were a free rotation the time average becomes an average over all azimuthal angles ϕ in fig. 44. We note that the function $\tfrac{1}{2}(3 \cos^2 \theta_{jk} - 1)$ is the second-degree Legendre polynomial $P_2(\cos \theta_{jk})$. This allows us to make use of the addition theorem for Legendre polynomials (see, for example, Margenau and Murphy (1943), p. 108, eq. 3-61) from which we find that the required average is

$$\overline{(3 \cos^2 \theta_{jk} - 1)}^{\phi} = 2\overline{P_2(\cos \theta_{jk})}^{\phi} = 2P_2(\cos \theta') P_2(\cos \gamma_{jk})$$
$$= \tfrac{1}{2}(3 \cos^2 \theta' - 1)(3 \cos^2 \gamma_{jk} - 1). \tag{6.20}$$

This average has been obtained for the case of free rotation. It can be shown however (Gutowsky and Pake, 1950) that it is also true for reorientation over, or tunnelling through, an n-fold periodic potential barrier, where $n \geqslant 3$. The spectrum therefore again consists of two lines

$$h = \pm \tfrac{3}{4}\mu r_{jk}^{-3}(3\cos^2\theta' - 1)(3\cos^2\gamma_{jk} - 1). \tag{6.21}$$

If the axis of reorientation ON is perpendicular to the internuclear vector OP, $(\gamma_{jk} = \tfrac{1}{2}\pi)$, as is commonly the case, this equation becomes

$$h = \pm \tfrac{3}{4}\mu r_{jk}^{-3}(3\cos^2\theta' - 1). \tag{6.22}$$

The dependence of the splitting on the orientation of the system with respect to $\mathbf{H}_0$ is thus similar in form to that given by (6.19), the axis of reorientation replacing the internuclear vector in specifying the orientation; the maximum splitting is, however, only half that for the rigid system. The similarity is more complete for a polycrystalline specimen, since random orientation of the crystal grains provides all values of either θ_{jk} or θ'. For a polycrystalline solid in which the internuclear vector of the isolated pair is effectively stationary at low temperatures, but reorients rapidly at higher temperatures about a perpendicular axis, the absorption line therefore undergoes a transition, becoming half as wide and correspondingly more intense at higher temperatures. This behaviour is exhibited by dichloroethane, $(CH_2Cl)_2$ (Gutowsky and Pake, 1950).

6.4.2. Systems of three identical nuclei. The case of triangular groups of identical nuclei of spin number $\tfrac{1}{2}$ reorienting about any given axis has been worked out by Andrew and Bersohn (1950). In general it leads, as in the rigid case discussed in §6.2.2, to a spectrum consisting of a central line and three pairs of lines symmetrically disposed about the centre. However, in the special case frequently encountered, in which the axis of reorientation is normal to the plane of the triangle, two pairs of lines disappear, and we are left with the central line, of weight $\tfrac{1}{2}$, and one pair of lines each of weight $\tfrac{1}{4}$. For a polycrystalline sample of acetonitrile, CH_3CN, the theoretical line shape is in agreement with that observed by Gutowsky and Pake (1950) at $93°$ K., and indicates that in this molecule the methyl group reorients about the C—C bond. Reorientation of the methyl group has now been established in this way in a large number of organic compounds (for references see Appendix 4).

6.4.3. More complicated systems. As we found with rigid

structures, it is difficult to calculate the line shape for reorienting systems containing more than three nuclei, and again we make use of Van Vleck's second-moment formula. We saw in §6.4.1 that for a dipole pair the factor $(3\cos^2\theta_{jk} - 1)$ had to be averaged over the motion. This suggests that in Van Vleck's formula (6.8) the terms U_{jk}, which contain this factor, must be averaged over whatever motion is occurring. That this is the correct procedure follows, as Gutowsky and Pake (1950) have shown, from the fact that the U_{jk} are coefficients which appear in the Hamiltonian expressing the magnetic interaction between the nuclei in the system.

Let us suppose that we have a single crystal containing a system of molecules or atomic groups each reorienting about one axis only, and suppose at first that the only nuclear dipoles present are those at resonance. We will divide the second moment S of the spectrum into an intramolecular part s_1 and an intermolecular part s_2. Then from (6.8) s_1 is given by

$$s_1 = \tfrac{3}{2}I(I+1)g^2\mu_0^2\mathcal{N}^{-1}\sum_{j>k}(\overline{U}_{jk})^2, \tag{6.23}$$

where $\overline{U}_{jk}$ is the average value of U_{jk} over the motion. For free rotation, or for reorientation over, or tunnelling through, an n-fold periodic potential barrier, where $n \geqslant 3$, we find $\overline{U}_{jk}$ from (6.9) and (6.20). Thus (6.23) becomes

$$s_1 = \tfrac{3}{8}I(I+1)g^2\mu_0^2\mathcal{N}^{-1}(3\cos^2\theta'-1)^2\sum_{j>k}(3\cos^2\gamma_{jk}-1)^2 r_{jk}^{-6}, \tag{6.24}$$

where, as before, θ' is the angle between the axis of reorientation and the applied field $\mathbf{H}_0$, and γ_{jk} is the angle between the internuclear vector $\mathbf{r}_{jk}$ and ON. If the molecule contains non-identical nuclear magnetic dipoles, the same averaging process must be applied to the intramolecular part of (6.11); the following additional term must then be added to (6.24):

$$\tfrac{1}{12}\mu_0^2\mathcal{N}^{-1}(3\cos^2\theta'-1)^2\sum_{j,f}I_f(I_f+1)g_f^2(3\cos^2\gamma_{jf}-1)^2 r_{jf}^{-6}, \tag{6.25}$$

where γ_{jf} is the angle between the internuclear vector $\mathbf{r}_{jf}$ and $\mathbf{H}_0$.

If the material is polycrystalline, there is an isotropic distribution of axes of reorientation, and the factor $(3\cos^2\theta'-1)^2$ occurring in (6.24) and (6.25) must be replaced by its mean value of $\tfrac{4}{5}$, giving

$$s_1 = \tfrac{3}{10}I(I+1)g^2\mu_0^2\mathcal{N}^{-1}\sum_{j>k}(3\cos^2\gamma_{jk}-1)^2 r_{jk}^{-6}$$
$$+ \tfrac{1}{15}\mu_0^2\mathcal{N}^{-1}\sum_{j,f}I_f(I_f+1)g_f^2(3\cos^2\gamma_{jf}-1)^2 r_{jf}^{-6}. \tag{6.26}$$

Comparison of (6.26) with equation (6.13) for a rigid structure shows that each term in the intramolecular contribution is reduced by a factor

$$\tfrac{1}{4}(3\cos^2\gamma_{jk}-1)^2. \tag{6.27}$$

This reduction factor decreases from unity (when $\gamma_{jk}=0°$) to zero (when $\gamma_{jk}=54°\,44'$) and increases again to $\tfrac{1}{4}$ (when $\gamma_{jk}=90°$).†

This reduction factor has been verified directly in one case by Andrew and Eades (1953 a, c). The benzene molecule reorients about its hexad axis in the crystal lattice, and causes the spectrum to narrow in the temperature range 90 to 120° K. The intramolecular contribution s_1 was found both below 90° K. and above 120° K. by the method of isotopic substitution discussed in §6.3; the best values were respectively 3·10 ±0·13 gauss² and 0·78 ±0·05 gauss². The ratio of these values is 0·252 ±0·02, in good agreement with the factor $\tfrac{1}{4}$ predicted by (6.27), since γ_{jk} is 90° for all internuclear pairs.

We thus see that the reduction in s_1 caused by molecular reorientation may be calculated quite simply. The reduction for the intermolecular contribution s_2 is more complicated, since in forming $\overline{U}_{jk}$ we have to remember that r_{jk} varies as well as θ_{jk}. Certain special cases have been treated by Andrew (1950), and formulae for the general case have been given by Andrew and Eades (1953 b); the application of these formulae usually involves a considerable amount of computing.

The transition, to which we have just referred, in the spectrum

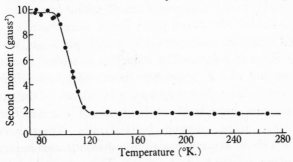

Fig. 45. Variation with temperature of the second moment of the proton magnetic resonance absorption line for polycrystalline benzene C_6H_6 (Andrew and Eades, 1953 c).

† An analysis of the reduction of s_1 by reorientation about more than one axis is given by Powles and Gutowsky (1953 a).

for benzene, C_6H_6, is illustrated in fig. 45, where the second moment is seen to decrease from a constant value of 9·7 gauss² below 90° K. to a constant value of 1·6 gauss² above 120° K. Fig. 46 shows the rather more interesting behaviour exhibited by poly-crystalline cyclohexane, C_6H_{12} (Andrew and Eades, 1952, 1953*b*). Below 150° K. the crystal structure is effectively rigid, and from the second moment it is possible to obtain a value for the HCH angle in the methylene groups of the molecule. Above 150° K. the

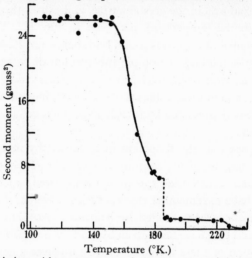

Fig. 46. Variation with temperature of the second moment of the proton magnetic resonance absorption line for polycrystalline cyclohexane C_6H_{12} (Andrew and Eades, 1953 *b*).

spectrum narrows, the second moment falling from 26·0 to 6·4 gauss² just below 186° K., at which temperature there is a change of crystal structure accompanied by a relatively large latent heat and expansion of volume. The reduction in second moment be-tween 150 and 180° K. appears to be caused by reorientation of the molecules about their triad axes; a calculation of the expected second moment, using the formulae discussed in this section, is in excellent agreement with the observed value. The polymorphic change at 186° K. is accompanied by a discontinuous reduction of second moment from 6·4 to 1·4 gauss². This value, which is main-tained up to 220° K., is quantitatively accounted for by assuming that the molecules reorient about axes other than their triad axes in the expanded lattice. This almost isotropic reorientation reduces the intramolecular contribution s_1 to zero, intramolecular local fields being averaged to zero. Local fields which are intermolecular in origin do not average to zero so long as the centres of mass of the molecules remain fixed. Between 220 and 240° K. even this remaining intermolecular contribution disappears. To account for this it is necessary to suppose that molecular centres of mass are no longer fixed and that the molecules diffuse through the lattice. This process is discussed further in §6.5.

Although we have discussed the characteristics of the spectrum both below and above the transition from an effectively rigid system to a rapidly reorienting system, we have not yet considered the spectrum in the process of transition. In the transition range the reorientation rate increases with temperature through values of the order 10^4–10^5 c./s. as we explained earlier. We may expect that the form of narrowing of the spectrum with increasing reorientation rate will follow the behaviour for liquids, given by (5.24), and in fact it is usually possible to fit the temperature variation of the width of the resonance line, such as is exhibited in figs. 45 and 46 to an equation based on (5.24) (Gutowsky and Pake, 1950; Andrew and Eades, 1953*b*, *c*).

As the table in Appendix 4 shows, molecular reorientation has now been established from nuclear magnetic resonance evidence in a great many compounds: in inorganic compounds such as ammonium salts, in almost spherical molecules such as methane and neopentane, in flat ones such as benzene, in long ones such as the long-chain paraffins and polymers, in liquid crystals, and in many others. It is important to note that materials investigated should be of the highest purity, since foreign molecules may deform the crystal lattice and thus assist, induce or prevent a reorientation process, or in other ways produce misleading results (see, for example, Rushworth, 1952).

6.5. Self-diffusion in solids

The resonance line for some solids has the extremely narrow width characteristic of liquids. An example, solid cyclohexane between 240 and 280° K., was mentioned at the end of the previous section. The most general form of reorientation of the molecules about fixed centres of mass merely reduces to zero the intramolecular contribution to the second moment, and leaves an inter-molecular contribution of one gauss^2 or so. The disappearance of this contribution also implies that the centres of mass are not fixed, and suggests that the molecules diffuse through the crystal lattice.

The self-diffusion of atoms in a metal is a well-established pheno-menon. It may, however, be thought that molecules, such as cyclo-hexane, which are some 6 Å. in diameter, are rather large and com-plex entities to move through a crystal lattice. Size is not in itself a relevant criterion however, and the following order of magnitude argument shows that the diffusion process is quite plausible. Self-diffusion is most likely to occur by the movement of molecules into neighbouring vacancies in the lattice; such vacancies are formed in increasing numbers as the melting-point is approached. Following Seitz (1951) the rate of jumping of a molecule into an adjacent vacancy may be written roughly as $\nu_v \exp(V/RT)$, where ν_v is the vibrational frequency of the molecule in the lattice, and V/N_0 is the sum of the energy required to form a vacancy in the lattice by removal of one molecule and the activation energy required to move an adjacent molecule into the vacancy (N_0 is Avogadro's number). If for solid cyclohexane the line width is to be affected by this diffusion process at 230° K., the rate of jumping must be of the order of the line width at this temperature (10^4 c./s.). The fre-quency ν_v is probably of the order 10^{12} c./s., though quite a large error in this value will not affect the conclusions which we shall draw. Taking this value of ν_v, it is found that V must be of the order of 8 kcal./mole for cyclohexane. For a vacancy mechanism V may be expected to have a value comparable with the lattice energy; certainly for cubic metals the value of V for self-diffusion is about two-thirds of the lattice energy. The lattice energy of cyclohexane is of the order of 10 kcal./mole; our value of 8 kcal./mole for V thus seems quite reasonable, and supports the suggested mech-anism of self-diffusion. Molecular solids such as cyclohexane are held together by weak van der Waals forces, with the result that although the cyclohexane molecule may be somewhat large, the lattice energy per molecule is less than the lattice energy per atom for many metals.

The first example of self-diffusion found by the nuclear mag-netic resonance method was that of solid hydrogen (Hatton and

Rollin, 1949). Besides the example of solid cyclohexane just mentioned, self-diffusion seems also to be indicated in solid methane (Thomas, Alpert and Torrey, 1950), solid methyl alcohol (Cooke and Drain, 1952), solid neopentane (Powles and Gutowsky, 1953 a) and solid cyclopentane (Rushworth, 1954). As we shall see in Chapter 7, self-diffusion in certain metals also has been indicated by a narrowing of the resonance line.

6.6. Spin-lattice relaxation in non-metallic solids

When dealing with liquids in Chapter 5, we found that the potent source of spin-lattice relaxation was the rapid variation of magnetic interaction between nuclear dipoles as the molecules which contained them performed their Brownian motion of rotation and diffusion. We have just encountered solids in which similar processes occur, and we may therefore expect the same mechanism of spin-lattice relaxation to be important for such solids. This we shall indeed find to be the case, but first we must consider the more usual type of solid in which molecular reorientation and translation are not found.

We consider then a crystal in which the magnetic nuclei are at fixed lattice sites, and whose only movement arises from the lattice vibrations appropriate to the temperature of the crystal. We will further assume for the present that all electronic configurations are diamagnetic, and that the magnetic nuclei have spin number $\frac{1}{2}$; by these assumptions we remove possible relaxation mechanisms involving paramagnetic ions and quadrupole interactions (the latter will be discussed in Chapter 8). For such a solid the most obvious mechanism to couple the nuclear spins to the lattice is the fluctuation of the local magnetic fields which arises from the lattice vibrations simply because the strength of the field at one nucleus due to the dipole moment of a neighbour depends on the distance between them. This mechanism was investigated by Waller (1932) in connexion with electronic paramagnetic relaxation, and the results may be applied to nuclear relaxation. He considered two processes. The first, or direct, process involves lattice vibrations at the nuclear magnetic resonance frequency ν_0, which enable the lattice to induce nuclear magnetic transitions, and thus to exchange energy with the spin system. This process turns out to be quite negligibly weak. The second process involves any pair of lattice modes which differ in frequency by the resonance frequency. The necessary non-linearity which mixes the two vibrational frequencies so that transitions occur at the beat frequency is the non-linearity of the term r^{-3} in the dipole interaction expression. This second process, though more effective than the first at ordinary temperatures, predicts relaxation times longer than those observed. At room temperature the predicted values of T_1 are at least 10^4 sec. and are often longer by several powers of ten, while as the temperature is lowered the predicted values increase as T^{-2} at first, and below the Debye temperature increase much more rapidly. Although T_1 is as long as 10^3 sec. for some solids at room temperature, it is usually found to be much shorter, while at low temperatures the discrepancy between measured and predicted values is many powers of ten. A related relaxation mechanism suggested by Purcell (1951) and Khutsishvili (1952) is no more successful than Waller's. Another related mechanism, in which the lattice vibrations are coupled to the nuclear spin system via electrons of the crystal's filled band, has been suggested by Muto and Watanabe (1952), but their calculations have been shown to be in error, and the mechanism is ineffective (Bloembergen, private communication). A more powerful relaxation mechanism has therefore to be found.

Such a mechanism has been suggested independently by Rollin and Hatton (1948, see also Hatton and Rollin, 1949), and Bloembergen (1949 a). This alternative mechanism arises from the fact that no actual solid is quite devoid of paramagnetic impurities, as we have so far supposed, and that either in the form of paramagnetic ions or F-centres they may play an important part in the relaxation process even at concentrations as low as one part per million. Bloembergen (1949 a) carried out a systematic investigation from 1 to 300° K. of the spin-lattice relaxation for protons in crystals of Al-alum in which a known small proportion of the Al^{3+} ions were replaced by paramagnetic Cr^{3+} ions, and found that T_1 was roughly inversely proportional to the paramagnetic ion concentration. He then developed a detailed mechanism which led to this observed dependence upon concentration, and which at all except the lowest temperatures gave fairly good absolute agreement with observed values of T_1. Any excess energy in the spin system diffuses by the mutual spin-exchange process described in §2.3 towards the large electronic magnetic moments of the impurity ions. The energy is then transferred to the lattice via the fluctuating magnetic field of these ions. The rate of disposal of excess energy by the diffusion process depends upon the concentration of impurity ions; but even if the concentration is only one part per million, there will on the average be a paramagnetic ion within about fifty lattice spacings of each proton. The transfer of energy to the lattice by the impurity ions depends upon the spin-lattice relaxation time for these ions; this time is, however, extremely short compared with nuclear spin-lattice relaxation times.

When the concentration of paramagnetic ions is greater than about 1%, the spin-diffusion process is unnecessary; within a few lattice spacings of every nuclear dipole there is an ion with which energy can be exchanged directly. Moreover, the magnetic interaction of the ions, now relatively close to each other, may endow their energy levels with a breadth greater than $h\nu_0$, the magnitude of the quanta which the nuclear dipoles wish to exchange. The system of ionic spins can now absorb these small quanta directly merely by rearrangement of the spins, a process which is independent of the lattice vibrations; the spin-lattice relaxation time is then short and independent of temperature, as Bloembergen (1949a) has found experimentally. In pure paramagnetic salts T_1 is very short even at $1°$ K.; for protons in hydrated copper sulphate Bloembergen (1949a) found T_1 less than 0·01 sec.

We now turn to solids containing reorienting molecules or groups. Such molecular motion usually provides the dominant relaxation mechanism except at the lowest temperatures. If, as we assumed in §6.4, the motion can be described by a single reorientation time τ_c, whose temperature-dependence is given by (6.15), then we expect T_1 to be given by equation (5.31). As we noticed in §5.4 this expression leads to values of T_1 falling as the temperature rises (falling values of τ_c), until a minimum is reached for $2\pi\nu_0\tau_c \sim 1$ (reorientation frequency and radiofrequency roughly equal), beyond which T_1 rises with temperature. Such behaviour has in fact been found with ammonium chloride and ammonium bromide in the solid state by Sachs and Turner (1951) and by Cooke and Drain (1952), and for three isotopic species of benzene by Andrew and Eades (1953 *c*); the experimental results for the latter case are shown in fig. 47.

Using (5.31), experimental values of T_1 may be converted into the corresponding values of τ_c. It is then possible to test the

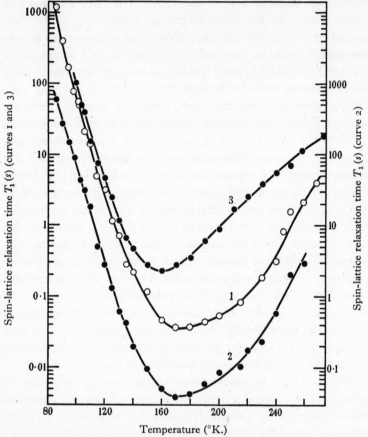

Fig. 47. Variation with temperature of the spin-lattice relaxation time for the protons in polycrystalline benzene. Curve 1, C_6H_6; curve 2, C_6H_5D; curve 3, 1.3.5-$C_6H_3D_3$ (Andrew and Eades, 1953 *c*).

assumed temperature-dependence of τ_c (equation (6.15)), by plotting log τ_c against $1/T$. In the cases just mentioned, of the ammonium halides and the benzenes, straight lines are obtained, covering a range of 10^6 in values of τ_c. For the benzenes these plots are shown in fig. 48. From the slopes of these lines the height V of the hindering potential barrier is found.

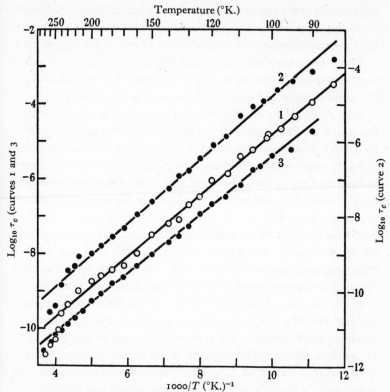

Fig. 48. Plot of $\log_{10} \tau_c$ against $1/T$ for polycrystalline benzene. Curve 1, C_6H_6; curve 2, C_6H_5D; curve 3, $1.3.5$-$C_6H_3D_3$; from the slopes of the straight lines the barrier heights were all found to be $3 \cdot 7 \pm 0 \cdot 2$ kcal./mole.

The diffusion of molecules in solids (§6.5) also provides a spin-lattice relaxation mechanism, as in the case of liquids. The effect of this mechanism is clearly noticeable in the case of solid cyclohexane (Andrew and Eades, 1953 b).

6.7. Special topics

6.7.1. The negative temperature experiment. Pound (1951) examined the ^{7}Li resonance in a very pure single crystal of lithium fluoride in a field of 6400 gauss at room temperature, and found a spin-lattice relaxation time T_1 of 300 sec., which is quite remarkably long at this temperature. It was found possible to remove the crystal from the magnet into the earth's field and return it after a few seconds with only a small loss of magnetization. A study of the dependence of the loss of magnetization (as indicated by the strength of the resonance signal) on the time the crystal was held out of the field indicated a relaxation time of about 15 sec. in the earth's field.

These facts enabled Purcell and Pound (1951) to carry out a most spectacular experiment with this crystal. The crystal, initially in equilibrium in the strong field of 6400 gauss, was quickly removed, through the earth's field, and placed in a small solenoid, the axis of which was parallel to a field of about 100 gauss, provided by a small permanent magnet. A 2 μF. condenser, initially charged to 8 kV., was discharged through the coil, with 500 ohms in series, so that the field in the coil reversed to $-$ 100 gauss in a very short time of about 0·2 μsec. and decayed back to the original field with the longer time constant of 1 msec. The initial rapid reversal of field took place in a time much less than a nuclear Larmor period (about 6 μsec. in 100 gauss), with the result that the nuclear magnetization was unable to follow the change and was thus opposed to the new sense of the magnetic field. The return to the original direction of the field took place in a time covering many Larmor periods; the magnetization was thus able to follow the changing field direction and remained opposed to the field. The crystal was then quickly returned to the strong magnet, and the ^{7}Li resonance inspected. The whole sequence of operations took only 3 sec. As fig. 49 shows, the resonance signal was negative corresponding to emission rather than absorption of radiofrequency power. This negative signal decayed through zero to the equilibrium state with the characteristic 300 sec. time constant.

When the nuclear magnetization has a direction opposed to that of the applied field, it follows that the upper energy levels are more populated than the lower levels. If it is granted that the system of nuclear spins may be properly described by a spin temperature (see §2.2), then this temperature must be negative. The first negative signal shown in fig. 49 therefore corresponds to a spin temperature of about $-$ 300° K. As the excess population of the upper levels falls, the spin temperature takes more negative values, reaching $-\infty$° K. when the levels are equally populated and the resonance signal strength is zero. As the lower levels become more populated, the temperature takes large positive values, falling from $+\infty$ to 300° K. One sees therefore that a negative temperature state is not cold, but is very hot, and gives up energy to any system at a positive temperature put in contact with it.

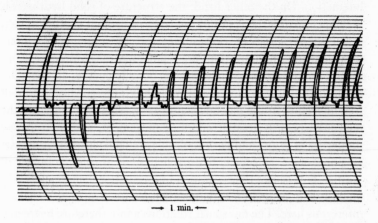

→ 1 min. ←

Fig. 49. The recovery of positive nuclear magnetization in LiF. On the left is a deflexion characteristic of the normal equilibrium magnetization at room temperature ($\sim 300°$K.), followed by a reversed deflexion corresponding to a spin temperature $\sim -350°$K., which decays through zero (infinite spin temperature) to the initial equilibrium state. Upward deflexions correspond to absorption and downward deflexions to emission (Purcell and Pound, 1951).

The long relaxation time of the lithium fluoride crystal allows it to be removed from a strong field and to be examined in weaker fields while it still possesses the relatively large magnetization corresponding to the strong field. Thus Ramsey and Pound (1951) were able to study the radiofrequency absorption spectrum in fields ranging from the earth's field strength to 42 gauss. Although internal fields caused the spectrum to cover continuously a frequency range of about 200 kc./s., the absorption was especially pronounced at the frequencies corresponding to the nuclear g-factors of ^{7}Li and ^{19}F. It would not of course have been possible to measure directly the absorption spectrum for such a solid in low fields, since, as the considerations of §3.9.1 show, the resonance absorption signal would have been much too weak. A combination of favourable parameters does, however, allow direct measurement of the absorption spectrum in zero applied field in the case of solid hydrogen; this is discussed in the following section.

6.7.2. Solid hydrogen. Solid hydrogen is a particularly interesting subject for investigation for a number of reasons. The interatomic distance in the hydrogen molecule is unusually short, 0·75 Å., giving rise to an especially large intramolecular proton interaction. On the other hand, the structure of solid hydrogen is very open, partly the result of the large zero-point energy; in fact, of all solids it has the lowest density. In consequence, intermolecular proton interactions are relatively weak. Moreover, since the molecules are extremely weakly bound together in the solid state, their characteristics are only slightly different from those of free molecules in the gaseous state. As we shall see presently the molecules possess great freedom of movement, and the separate properties of *ortho-* and *para-*molecules remain clearly distinct. Of all polyatomic molecules, hydrogen has the smallest moment of inertia. This causes the separation of its rotational energy levels to be unusually wide, in fact, larger than kT for temperatures which interest us here. The molecular rotation cannot therefore be treated classically.

Hatton and Rollin (1949) examined the proton magnetic resonance in solid hydrogen between its freezing-point (14° K.) and 1° K., using hydrogen having the room-temperature equilibrium *ortho-para* ratio of 3 : 1. The observed resonance comes solely from the *ortho-*molecules, whose resultant spin number $I = 1$, with rotational quantum number $J = 1$; the *para-*molecules with $I = 0$, $J = 0$ take no part in the resonance. They found that from 14° K. down to 11° K. the *ortho-*molecules underwent almost isotropic reorientation, and diffused rapidly from one lattice site to another, giving a very narrow resonance line (see §§6.4 and 6.5). Between 11 and 10° K. the line broadened to a width of several gauss, which was satisfactorily accounted for by supposing that the self-diffusion had ceased to be sufficiently rapid to affect the line width; the line width is thus purely intermolecular in origin in this temperature region. The line width then remained fairly constant down to about 1·5° K., below which it broadened out again. Reif and Purcell (1953) confirmed this behaviour in the region below 4·2° K., and accurately recorded the line shape, shown in fig. 50, at 1·16° K. The further substantial increase in width below 1·5° K. must be intramolecular in origin. If, however, the mechanism were to consist of a complete suppression of molecular rotation, then the resonance spectrum for the resulting rigid array of protons should be that discussed in §6.2.1 and exemplified in fig. 41. The observed spectrum (fig. 50) is indeed of this shape, but the separation between the peaks is only about 40% of that indicated by equation (6.6). This equation indicates a separation of $3\mu r^{-3}$, which has the

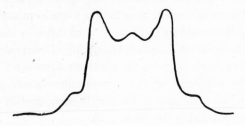

Fig. 50. Schematic representation of the observed proton magnetic resonance line shape in solid hydrogen (Reif and Purcell, 1953).

unusually large value of 100 gauss for hydrogen, since the internuclear distance r is so small. It is clear therefore that molecular rotation is not entirely suppressed, and that instead an asymmetry of the intermolecular forces prevents each molecule from taking all directions with equal probability.

For isolated fixed molecules whose internuclear vector makes an angle θ with respect to the direction of the applied magnetic field $\mathbf{H}_0$, the spectrum consists of two lines given by (6.5):

$$h = \pm\tfrac{3}{2}\mu r^{-3}(3\cos^2\theta - 1). \tag{6.28}$$

The considerations of §6.4 show that in a reorienting system the angular factor $(3\cos^2\theta - 1)$ must be replaced by its time average. In the cases discussed in §6.4, reorientation could usually be assumed to take place about a well-defined axis, and a straightforward classical average of $(3\cos^2\theta - 1)$ could be taken. Here, the departure of the potential, in which the molecule moves, from spherical symmetry, is small compared with the energy of the molecule, and can thus be treated as a small perturbation of that for a free rotator. Reif and Purcell (1953) have solved the perturbation problem and have found the angular probability distribution for the internuclear vector r. Using this, the average value of $(3\cos^2\theta - 1)$ is found to be $\tfrac{2}{5}(3\cos^2\epsilon - 1)$, where ϵ is the angle between $\mathbf{H}_0$ and a principal axis of the asymmetric potential in which a given molecule finds itself. For molecules with the same ϵ, the spectrum is thus

$$h = \pm \tfrac{3}{2}\mu r^{-3} . \tfrac{2}{5}(3\cos^2\epsilon - 1). \qquad (6.29)$$

The spectrum for a polycrystalline aggregate containing an isotropic distribution of angles ϵ thus has the same form as that shown in fig. 41 and described by (6.6), but scaled down in width by the factor $\tfrac{2}{5}$ by which the coefficient in (6.29) differs from that in (6.28). This theoretical line shape is in excellent agreement with that observed. The relatively small intermolecular broadening allows the details of the structure to be clearly seen, and in particular shows the steps flanking the main peaks.

Reif and Purcell (1953) were also able to measure directly, in the absence of an external field $\mathbf{H}_0$, the dipole-dipole interaction responsible for the hydrogen resonance in high fields below $1\cdot5^\circ$ K. In principle the detection of zero-field resonances of this kind would also be possible in other cases, as, for example, for the protons in gypsum, whose spectrum we discussed in §6.2.1. Solid hydrogen is particularly favourable for such an experiment however, since an adequate signal-to-noise ratio is assured by the large dipole-dipole interaction (small r), the small intermolecular broadening, a short spin-lattice relaxation time, and the low temperature. It turned out from a theoretical study that the zero-field resonance frequency should be about 165 kc./s., equal to the interval between the peaks in the high field resonance line, when converted into frequency units. A relatively large sample, 8 cm.^{3}, of solid hydrogen was used. The resonance was observed at the expected frequency with a width of about 25 kc./s. determined by intermolecular dipole interactions. This experiment provided a valuable check on the value of μr^{-3} found from the high field experiments and also of the theoretical considerations upon which the interpretation of the high field experiment was based.

6.7.3. Ferroelectrics. Newman (1950 a) has examined crystals of potassium dihydrogen phosphate, KH_2PO_4, and potassium dihydrogen arsenate, KH_2AsO_4, by the nuclear magnetic resonance method. These salts exhibit spontaneous electric polarization below a certain temperature (122° K. for KH_2PO_4 and 96° K. for KH_2AsO_4), in a manner analogous to the magnetic behaviour of ferromagnetics. The hydrogen atoms in KH_2PO_4 form hydrogen bonds between oxygen atoms on the corners of adjacent PO_4 tetrahedra. Slater (1941) in his theoretical approach supposes that the protons are not placed at the centre of the hydrogen bonds, but take up a position closer to one or other oxygen atom. Either position is possible subject to the proviso that each PO_4 tetrahedron shall have two close protons and two remote protons situated on its four hydrogen bonds. It is thus supposed that each proton may move from one position to the other, over or through any intervening barrier, to enable the crystal as a whole to take up its most favourable configuration. If such motion were sufficiently rapid we should expect the resonance line to be narrowed, and we should expect the motion to provide a source of spin-lattice relaxation.

Newman did in fact observe a narrowing of the resonance line at about $240°$ K. for KH_2PO_4, and at about $200°$ K. for KH_2AsO_4, which is consistent with the view of protons shuttling back and forth at a rate which has reached about 10 kc./s. at these temperatures (well above the ferroelectric region of course). He found that T_1 varies only slowly with temperature for both salts. This insensitivity to temperature led him to suggest that the protons may be tunnelling through the potential barrier near its base rather than surmounting it.

6.7.4. Paramagnetics and antiferromagnetics.

So far in this chapter we have been almost exclusively concerned with solids which do not contain paramagnetic ions, or which contain only a small concentration of such ions. Two paramagnetic salts have been studied in some detail. Bloembergen (1950) has examined the proton resonance in monocrystalline copper sulphate pentahydrate, $CuSO_4.5H_2O$, both theoretically and experimentally. This work was continued by Poulis (1951), who also examined the proton resonance in monocrystalline cupric chloride dihydrate, $CuCl_2.2H_2O$.

In these crystals each proton has at least one paramagnetic cupric ion as a close neighbour, and this has two main effects. In the first place, as we have already discussed in §6.6, the presence of the cupric ions markedly promotes the spin-lattice relaxation process for the protons. Secondly, the local magnetic field of the cupric ions causes the resonance spectrum to be very broad, especially at low temperatures.

The magnetic moments of the cupric ions are continually changing their orientation by exchanging quanta of energy with each other and with the lattice, in much the same way as do the nuclear magnetic moments. However, since the magnetic moment of an ion is about a thousand times larger than a nuclear magnetic moment, the reorientation of the ionic moment is a very much more rapid process. Nevertheless, this rapid reorientation does not mean that the average ionic moment is zero, since, unlike the case of nuclear moments, the separation of energy levels in an applied field of several kilogauss is not very small compared with kT; the appreciably greater numbers of ions in the lower states results in a finite average moment. In fact, the time-averaged magnetic moment $\bar{\mu}_c$ of the cupric ions is given in order of magnitude by Curie's law

$$\bar{\mu}_c = \frac{\mu_c^2 H_0}{3kT}, \qquad (6.30)$$

where μ_c is the magnetic moment of the cupric ion. In an applied field of 10 kilogauss at 300° K., the value of $\bar{\mu}_c$ given by (6.30) is of the order of 6 nuclear magnetons, and at low temperatures is much larger. This average magnetization is responsible for a local magnetic field which varies strongly, of course, with the space coordinates over the unit cell, so that the different protons in the unit cell have different resonance frequencies. At low temperatures the resonance therefore splits into a number of component lines, and the order of magnitude of the splitting is given by $\bar{\mu}_c/r^3$, where r is the distance between a proton and a copper nucleus. With $r \sim 2.5$ Å., the splitting expected is ~ 2 gauss at 300° K., ~ 30 gauss at 20° K. and ~ 600 gauss at 1° K. The triclinic unit cell of copper sulphate contains two cupric ions and hence twenty protons. Since the unit cell possesses a centre of symmetry, the number of protons having a different environment is ten, and we therefore expect the resonance to be split into ten lines. Bloembergen (1950) did in fact find ten lines at temperatures of 4° K. and below. At higher temperatures where the splitting is smaller the lines were not all resolved. One may regard the protons as small magnetic indicators disposed throughout the unit cell, which by their individual resonance frequencies register the local field of the paramagnetic ions at various points in the unit cell. Poulis (1951) observed for copper sulphate that the splitting of the resonance lines did follow very roughly the inverse temperature-dependence predicted by the Curie law, except at the lowest temperatures.

Since each proton has one close proton neighbour, namely, its partner in the same water molecule, each of the ten lines is actually a doublet. Poulis (1951) was able to resolve some of these doublets at $1\cdot5^\circ$ K., but he was never able to obtain all twenty lines, because for every orientation of the crystal some of them overlapped. He therefore turned to an investigation of $CuCl_2.2H_2O$ which has a simpler structure. The orthorhombic unit cell contains four different protons, and in general four resonance doublets could be resolved with this crystal. From room temperature down to 14° K. the crystal behaved in the expected manner. Assuming that the mean magnetic moment $\overline{\mu}_c$ of a cupric ion is directed parallel to $\mathbf{H}_0$, the component H_c of local field parallel to $\mathbf{H}_0$ at a given proton due to its nearest cupric ion neighbour is

$$H_c = \overline{\mu}_c (3 \cos^2 \theta_c - 1) r_c^{-3}, \qquad (6.31)$$

where $\mathbf{r}_c$ is the vector joining the ion to the proton, and θ_c is the angle between $\mathbf{H}_0$ and $\mathbf{r}_c$. As the crystal is turned about a given axis the local field, and therefore also the resonance shift, due to the nearest cupric ion should vary sinusoidally with period 180°. Fig. $51\,a$ shows that this is approximately the case at $14\cdot3^\circ$ K. for the four resonance lines obtained when the crystal is turned about the c axis with $\mathbf{H}_0$ in the ab plane. The value of $\overline{\mu}_c$ obtained from the amplitude of the sinusoidal curves is a reasonably good approximation to the Curie law (6.30) down to 14° K., provided one makes the Weiss substitution of $(T + T_0)$ for T, with T_0 about 4° K.

At $4\cdot3^\circ$ K. a remarkable change of behaviour took place which Poulis and Hardeman (1952 a, b) were able to explain by assuming that the salt became antiferromagnetic below this temperature.[†] Susceptibility measurements have since given direct confirmation of this assumption. The evidence suggests that the cupric ions are no longer free to reorientate, but that in weak magnetic fields the

[†] An earlier attempt to investigate the antiferromagnetic crystal MnF_2 using the ^{19}F resonance had been unsuccessful (Bloembergen and Poulis, 1950).

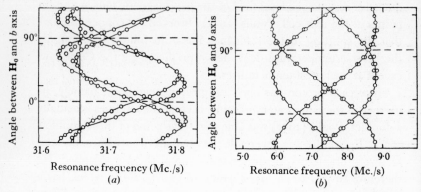

Fig. 51. Resonance frequencies of the four proton lines from a single crystal of $CuCl_2.2H_2O$ as the crystal is turned about its c axis with the applied field $\mathbf{H_0}$ in the ab plane (Poulis and Hardeman, 1952 a). For (a) the field strength was about 7500 gauss (resonance frequency for protons in water 31·66 Mc./s.) and the temperature was 14·3 °K.; for (b) the field strength was about 1700 gauss (7·26 Mc./s.) and the temperature 4·13 °K. Note that in (a) only four resonance lines (two doublets) are observed instead of the four doublets found in more general crystal orientations; four partially resolved doublets are however found in (b) for a reason discussed in the text.

crystal consists of two interpenetrating lattices with antiparallel cupric ion spins, the spins being directed along the $\pm a$ axes. The cessation of rapid ionic reorientation removes the principal source of proton spin-lattice relaxation, and in consequence the resonance signals are much weaker below than above 4·3° K.

Fig. 51 b shows the angular variation of the resonance frequencies at 4·13° K. in a weak field (1700 gauss), for the same orientations of the crystal as in fig. 51 a. The following differences between the figures are apparent:

(a) The curves in fig. 51 b are almost symmetrical about the resonance frequency for a free proton. Thus for every ionic magnetic moment oriented in a given direction there must be another oppositely directed. This also explains why there are four doublet resonances at each orientation instead of two doublets in fig. 51 a.

(b) The periodicity of angular variation is 360° rather than 180°. Since the cupric ion magnetic moment is now directed along the $\pm a$ axes instead of parallel to $\mathbf{H_0}$, the local field component due to a given ion is

$$H_c = \pm\mu_c(3\cos\theta_c\cos\Psi - 1)r_c^{-3}, \qquad (6.32)$$

where Ψ is the angle between $\mathbf{H}_0$ and the a axis; this expression requires a periodicity of 360°.

The magnetic moment of the cupric ion, as indicated by the amplitude of the sinusoidal excursions in diagrams such as fig. 51b, does not of course follow a Curie–Weiss law in this region, and has the variation shown in fig. 52 (Poulis and Hardeman, 1952a,

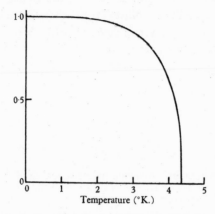

Fig. 52. The temperature-dependence of the magnetic moment of the cupric ion in the antiferromagnetic state of $CuCl_2 \cdot 2H_2O$ (Poulis and Hardeman, 1953a). The ordinate gives relative values of the ionic magnetic moment.

1953a). Poulis, Hardeman and Bölger (1952) find that the Néel temperature deduced from their proton resonance measurements depends somewhat upon the direction of the applied field $\mathbf{H}_0$; for a weak field parallel to the a axis the Néel temperature is 4·293° K., whilst when the field is parallel to the b axis it is 4·337° K.

In stronger magnetic fields of 5 kilogauss or more, the anti-parallel arrangement of ionic spins along the a axis is disturbed, and the angle by which the spins deviate from the a direction depends upon the direction and strength of the applied field. In fields of the order of 10 kilogauss the ionic spins tend to set themselves perpendicular to the applied field. The complicated angular variation of proton resonance lines in this range of magnetic field has been studied by Poulis and Hardeman (1952a, b, 1954), and their results give valuable information concerning the antiferro-magnetic behaviour of the cupric ions. Gorter (1953) gives a review of four related studies of this interesting salt involving measurements of static magnetization, specific heat, proton magnetic resonance and antiferromagnetic resonance.

Zener (1951) has suggested the existence of antiferromagnetic coupling of $3d$ electrons in a number of transition metals such as vanadium which have body-centred cubic crystal structures. The resonance spectrum for these metals (see § 7.4) does not have the considerable breadth found for the antiferromagnetic salt discussed above. Knight and Kittel (1952) therefore conclude that if in these metals an antiferromagnetic array of spins of the $3d$ electrons exists it is not a static array.

Note added in proof

Several important papers on nuclear magnetic resonance in solids were read at the Bristol Conference on Defects in Crystalline Solids, July 1954, published by the Physical Society (1955).

CHAPTER 7

NUCLEAR MAGNETIC RESONANCE IN METALS

7.1. Introduction

There are two main differences in behaviour between metals and non-metals, and both arise from the presence of the conduction electrons.

The most important difference is a practical one and arises from the skin effect. The radiofrequency magnetic field is able to penetrate into a metallic surface only to a depth of the order of 5×10^{-3} cm. in typical experiments. If large pieces of metal are used as specimens only those nuclei in a thin surface layer of this depth are therefore able to participate in the nuclear magnetic resonance.

Despite the loss of signal strength entailed by the skin effect, Pound (1948 *b*) was able to discover the resonances of ^{63}Cu and ^{65}Cu in the copper wire of which his specimen coil was made. Later workers have usually improved the signal strength by the use of finely divided metal which presents a greater surface for the same volume of material. Some workers have used thin foils, others powders dispersed in paraffin wax, while others have dispersed metals of low melting-point in oil by melting them and stirring at high speed or applying ultrasonic vibration.

When the specimen dimension is small compared with the skin depth, the radiofrequency field penetrates the whole specimen with practically no attenuation, and the situation is not different from an insulator. If, however, the dimensions are not small compared with the skin depth the situations do differ. The change in magnetic susceptibility at resonance is accompanied by a change in skin depth, and consequently the conduction losses are also changed. The total absorption is then determined by a combination of both real and imaginary parts of the complex susceptibility. The magnetic resonance absorption line may then become shifted and distorted. A full analysis of this effect has been given by Bloembergen (1952). If one intends to make a careful study of the resonance line shape and the resonance frequency it is therefore important to use specimens whose narrow dimension is less than the skin depth.

The preparation of small specimens may introduce other uncertainties however. Impurities may be picked up in the process of subdivision, and operations of rolling, filing and grinding are likely to leave the metal in a strained condition. Unless precautions are taken, the specimen may therefore not be representative of pure bulk metal.

The second main difference in behaviour between metals and non-metals arises from the magnetic effects of the metallic conduction electrons. They are responsible for spin-lattice relaxation, as will be discussed in the next section, and they also give rise to a resonance frequency shift to be discussed in §7.3.

A list of metals which have been investigated experimentally is given, with references, in Appendix 5.

7.2. Spin-lattice relaxation time for metals

Several spin-lattice relaxation mechanisms which are effective in insulators can also be effective in metals: (a) paramagnetic ion impurity mechanism (see §6.6); (b) for $I > \frac{1}{2}$, relaxation effected through electric interaction between the lattice and the nuclear electric quadrupole moment (see §8.4); (c) relaxation effected by diffusion of metal atoms (see §6.6). However, the most potent source of spin-lattice relaxation in a metal appears, from the studies made so far, to be that provided by the conduction electrons.

When an electron passes close to a nucleus, the nucleus experiences a relatively large time-varying local magnetic field, which may induce transitions between the magnetic energy sub-levels of the nucleus. The energy $h\nu$ emitted or absorbed by the nucleus is taken up or surrendered by the electron by adjustment of its kinetic energy. Transitions are therefore only possible if there is a vacant level available for the electron which differs in energy by an amount $h\nu$ from its initial energy. This requirement renders the relaxation process considerably less powerful than one might at first think, since in a Fermi–Dirac distribution vacancies only occur in a range of energy whose width is of order kT at the top of the energy distribution. Since $h\nu \ll kT$ in all practical cases, only a fraction of order kT/E_0 of the electrons are able to participate in the relaxation process, where E_0 is the energy at the top of the Fermi distribution. Even at room temperature this fraction will usually be only of the order of 1%.

Heitler and Teller (1936) showed by an order of magnitude argument that the probability W of a nuclear transition being induced by conduction electrons is

$$W \cong \frac{E_1^2 kT}{E_0^2 h} , \qquad (7.1)$$

where E_1 is an energy of the order of the hyperfine splitting of the ground state of the metal atom. Using equation (2.11) we see that expression (7.1) is just $1/2T_1$, where, as usual, T_1 is the spin-lattice relaxation time. At room temperature this expression leads to values of T_1 of the order of 10^{-2} sec., while even at $1°$ K. the values are only of the order of 1 sec., notwithstanding the fact that only a proportion kT/E_0 of the electrons are effective in the relaxation process. We are in fact not surprised to see this factor appearing in (7.1), and bringing about an inverse proportionality between T_1 and the absolute temperature.

Expression (7.1) may be derived by the following order of magnitude argument, based on that given by Bloembergen (1949 b). Most metals contain approximately one conduction electron per atom. There is therefore, on the average, about one conduction electron in the vicinity of each nucleus, that is to say, within a distance of the order of the crystal-lattice spacing r_0. If an electron moves with velocity v it spends a time $t \sim r_0/v$ in the vicinity of a given nucleus. If, therefore, all conduction electrons moved with velocity v and each exchanged a quantum of energy with the nucleus while in its vicinity, the rate at which exchanges are made would be $1/t$. However, as we mentioned, only a fraction kT/E_0 of electrons are effective in the relaxation process. If all these effective electrons did exchange a quantum, the rate at which exchanges are made would be

$$\frac{kT}{E_0 t} \cong \frac{kTv}{E_0 r_0} , \qquad (7.2)$$

where v must now refer to the velocity of the electrons at the top of the Fermi distribution. Actually, however, only a small fraction of the encounters between an effective electron and a nucleus produce an exchange of energy. From perturbation theory the probability of such an exchange is

$$\hbar^{-2} \mid V_{ij} \mid^2 t^2, \tag{7.3}$$

where V_{ij} is the matrix element of the magnetic interaction between the electronic and nuclear spins, the initial and final states of the nucleus being described by magnetic quantum numbers m_i and m_j respectively. The order of magnitude of this matrix element will be given by the magnetic interaction energy

$$\mid V_{ij} \mid \sim E_1 \simeq \frac{\mu_1 \mu_2}{r_{12}^3}, \tag{7.4}$$

where μ_1 and μ_2 are respectively the magnetic moments of the nucleus and the electron, and r_{12} is their distance of closest approach. The energy E_1, and therefore $\mid V_{ij} \mid$ also, will thus be of the order of the hyperfine splitting of the ground state of the atom. The transition probability W is therefore given by the product of (7.2) and (7.3), which, using (7.4), gives

$$W = \frac{kTE_1^2 t}{E_0 \hbar^2} = \left(\frac{E_1^2 kT}{E_0^2 \hbar^2} \right) (\tfrac{1}{2} E_0 M_e r_0^2)^{\frac{1}{2}}, \tag{7.5}$$

since $t \sim r_0/v$ and $v = (2E_0/M_e)^{\frac{1}{2}}$, where M_e is the electronic mass and E_0 is the well-known degeneracy parameter given by

$$E_0 = (3\pi^2)^{\frac{2}{3}} \left(\frac{\hbar^2}{2M_e r_0^2} \right). \tag{7.6}$$

Using this relation in the last factor of (7.5) we get

$$W = \frac{1}{2T_1} = \left(\frac{3\pi^2}{8} \right)^{\frac{1}{3}} \frac{E_1^2 kT}{E_0^2 \hbar}, \tag{7.7}$$

which, apart from a numerical factor of order unity, is the expression given in (7.1). A more rigorous derivation is given by Korringa (1950).

Rollin and Hatton (1948) made the first measurement of T_1 for a metal, and for ^{27}Al found a value of about 1 second between 1 and 4° K. This is in order of magnitude agreement with (7.1). Bloembergen (1949 b) made a more systematic study of the dependence of T_1 upon temperature and verified the predicted inverse dependence for ^{63}Cu and ^{65}Cu. His results are shown in fig. 53, and it will be seen that the values of T_1 for ^{63}Cu are some 15% higher than those for ^{65}Cu. This accords roughly with the proportionality between T_1 and E_1^{-2} (see equation (7.7)); since E_1 is proportional to the nuclear magnetic moment, and since the ratio of the moment for

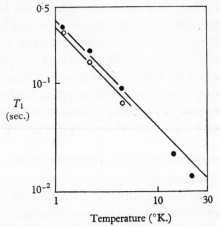

Fig. 53. The temperature variation of the spin-lattice relaxation time for ^{63}Cu and ^{65}Cu in metallic copper at 9·5 Mc./s. (Bloembergen, 1949 b). Closed circles ^{63}Cu; open circles, ^{65}Cu.

^{65}Cu to that of ^{63}Cu is 1·07, the value of T_1 for ^{63}Cu should indeed be some 15% higher than that for ^{65}Cu. There is, however, an unexplained discrepancy between the value of 0·3 sec. found at 1·2° K. by Bloembergen (1949 b) and the value of 3 sec. found by Hatton and Rollin (1949) at the same temperature.

The inverse temperature dependence has been confirmed also for ^{23}Na between 215 and 353° K. (Norberg and Slichter, 1951) and for ^{27}Al between 1·2 and 4·2° K. (Poulis, 1950). For ^{7}Li, however, Poulis (1950) found deviations from the inverse temperature-dependence. At 1° K. the relaxation time is about 20 sec. for this metal, which is of the order of a hundred times longer than for aluminium and copper. It is possible, therefore, as Korringa (1950) points out, that the deviations are caused by the presence of a small concentration of paramagnetic impurity. As we saw in §6.6 such impurities can provide an important relaxation mechanism, and their effect would be more noticeable when, as with ^{7}Li, the conduction electron relaxation mechanism is relatively weak.

Overhauser (1953) has shown that the application of saturating radiofrequency power to a metal in a magnetic field at the electron-spin resonance frequency (Griswold, Kip and Kittel, 1952) produces a polarization of the nuclear spins in the metal, as a consequence of the interaction between electrons and nuclei which we have been discussing. The excess of nuclei in the lower nuclear magnetic energy levels can in this way be increased by a factor of order 10^3, and the nuclear magnetic resonance signal is correspondingly strengthened. Carver and Slichter (1953) subjected a specimen of lithium in a field of 30 gauss to a saturating power of 84 Mc./s. (the electron-spin resonance frequency), and found a strong ^{7}Li nuclear resonance at 50 kc./s.; in absence of the electron-spin resonance radiation the nuclear resonance was unobservably weak.

7.3. Nuclear magnetic resonance shift for metals

Knight (1949) noticed that the nuclear magnetic resonance signal from a metallic specimen occurred at a higher field strength than that for the same nuclear species in a non-metallic specimen. The shift ΔH_0 was found to be proportional to the applied field H_0. The effect has now been found for about ten metals and appears to be general. The shift shows a general increase with atomic number, the ratio $\Delta H_0/H_0$ rising from 0.03×10^{-3} for ^{7}Li (Knight, 1949) to 12×10^{-3} for ^{207}Pb (Townes, Herring and Knight, 1950).

The shifts are much too large to be accounted for by a simple difference in magnetic susceptibility of the materials, or by differences in diamagnetic correction for the metallic and non-metallic atoms (see §4.1). It has therefore been suggested by Townes (Knight, 1949; Townes, Herring and Knight, 1950) that the shifts are caused by the magnetic interaction with the nuclei of the conduction electrons near the top of the Fermi distribution. Since this interaction is also responsible for the spin-lattice relaxation (§7.2) we may expect a relationship to exist between the relaxation time and the shift. Such a relation will be discussed later in this section.

Although the operation of Fermi–Dirac statistics causes the macroscopic average susceptibility of a metal to be very small, the local magnetic susceptibility is very much greater since the electrons have a large probability density near the nucleus. The shift ΔH_0 is just the increase in mean magnetic field at the nucleus caused by the conduction electrons and, assuming cubic symmetry, is given by $\frac{8}{3}\pi$ times the mean density of electron magnetic moment at the nucleus (i.e. the local magnetic susceptibility multiplied by H_0). If the electron distribution does not have cubic symmetry about the nucleus the shift is anisotropic.

If χ_e is the contribution of the conduction electrons to the macroscopic susceptibility per unit volume, the mean density of magnetic moment at a given nucleus is

$$\frac{\chi_e H_0}{N} \overline{|\psi_F(0)|^2}, \qquad (7.8)$$

where N is the number of nuclei per unit volume, and $\overline{|\psi_F(0)|^2}$ is the average probability density at the nucleus for all electronic states on the Fermi surface, expressed in terms of the wave function of the electrons. Hence, applying the factor $\frac{8}{3}\pi$, we find

$$\frac{\Delta H_0}{H_0} = \frac{8\pi\chi_e}{3N} \overline{|\psi_F(0)|^2}. \qquad (7.9)$$

If the wave functions ψ_F were completely known, equation (7.9) would enable the theoretical shift to be calculated directly and compared with experiment. Since this is not the case, one recalls that the interaction between electron and nucleus is of the same form as that which produces the hyperfine splitting in a free atom. For an s electron Fermi (1930) has shown that this splitting, expressed as an energy, is given by

$$E_1 = \frac{8\pi(2I+1)\mu\beta}{3I} |\psi_a(0)|^2, \qquad (7.10)$$

where, as usual, μ and I are the magnetic moment and spin number of the nucleus, β is the Bohr magneton, and where ψ_a is the wave function for a free atom. Combining (7.9) and (7.10) we have

$$\frac{\Delta H_0}{H_0} = \frac{E_1 \chi_e I \Gamma}{\mu\beta N(2I+1)}, \qquad (7.11)$$

where

$$\Gamma = \frac{\overline{|\psi_F(0)|^2}}{|\psi_a(0)|^2}. \qquad (7.12)$$

The factor Γ is of the order of unity and has been worked out for lithium, beryllium and sodium (Townes, Herring and Knight, 1950; Kohn and Bloembergen, 1950; Jones and Schiff, 1954). Using these values of Γ, and known values of E_1 and χ_e, the expected shift is calculated from (7.11), and is found to be in fair agreement with observed values of $\Delta H_0/H_0$. The agreement is in fact good enough to support the proposed mechanism of the shift, and the failure to get exact agreement probably arises from the lack of accurate values for χ_e, and of accurate wave functions ψ_F. Indeed, the shift can provide information concerning the nature of the electronic states in metals at the top of the Fermi distribution (Jones and Schiff, 1954).

The observed increase of the shift with atomic number is not unexpected since, as we see from (7.11), the shift is proportional to the hyperfine splitting E_1 of the free atom, which is well known to increase with atomic number.

We now return to the point mentioned earlier in this section, that there should be a relationship between the shift and the spin-lattice relaxation mechanism, since both originate in the interaction between the nuclei and the conduction electrons. First we substitute in (7.11) the well-known approximate expression for χ_e given by the free-electron gas model (see, for example, Mott and Jones, 1936, p. 186):

$$\chi_e = \tfrac{3}{2}N\beta^2/E_0. \tag{7.13}$$

Substituting this in (7.11) we get

$$\frac{\Delta H_0}{H_0} = \frac{3\beta I \Gamma}{2\mu(2I+1)}\frac{E_1}{E_0}. \tag{7.14}$$

From (7.1) and (2.11) we have

$$T_1 = \frac{hE_0^2}{2kTE_1^2}. \tag{7.15}$$

Eliminating from (7.14) and (7.15) the energy ratio (E_1/E_0), we obtain

$$T_1\left(\frac{\Delta H_0}{H_0}\right)^2 = \frac{\hbar G}{\pi kT}\left(\frac{\beta I}{\mu}\right)^2, \tag{7.16}$$

where G is a numerical factor of order unity. This approximate relation between the spin-lattice relaxation time and the resonance shift has been derived more rigorously by Korringa (1950). He has tested the relation for lithium, aluminium and copper for which experimental values of both T_1 and $\Delta H_0/H_0$ have been obtained, and the two sides of (7.16) are found to be in rough agreement.

Most of the metals investigated have a cubic structure, but we must mention the anisotropy of the resonance shift which is to be expected in non-cubic metals. Since it is usually necessary to work with finely divided material in order to get adequate signal strength, the specimen inevitably contains crystallites of all orientations. If the shift is anisotropic the resonance line is therefore broadened. Bloembergen and Rowland (1953) were able to account for the asymmetrically broadened resonance they found with white tin, which has tetragonal symmetry, by assuming a shift $\Delta H_0/H_0$ of $7\cdot9 \times 10^{-3}$ along the tetragonal axis and $7\cdot4 \times 10^{-3}$ perpendicular to the axis.

The resonance shift has been found to be relatively insensitive to temperature (Gutowsky, 1951; McGarvey and Gutowsky, 1953), and such small variation as was found can be accounted for in terms of expansion of the crystal lattice.

7.4. Nuclear magnetic resonance absorption spectrum for metals

There are four important sources of broadening of the resonance absorption line for metals. They are (a) nuclear magnetic dipolar interaction, (b) nuclear electric quadrupole interaction, (c) broadening associated with a short spin-lattice relaxation time (see §5.3), (d) anisotropy of the resonance shift.†

Of these four sources of broadening, source (a) is always present. Source (b) is only present when there is a crystalline electric field gradient with which the nuclear quadrupole moment can interact, and is therefore only found in metals with non-cubic crystal structure. Source (c) is always present and frequently is of importance. Source (d) is peculiar to metals, and was discussed in §7.3; it is to be found only in non-cubic metals.

We therefore begin our account with cubic metals, a group which includes the majority of metals studied so far. For these metals sources (b) and (d) should be absent. At 77° K. the spin-lattice relaxation times for lithium (^{7}Li) and sodium (^{23}Na) are too

† A further source of broadening is produced by the exchange coupling of nuclear spins by the conduction electrons (Ruderman and Kittel, 1954). This broadening, which is very similar to that discussed for molecules in §5.7, appears to be responsible for the anomalous breadth of the resonance lines in metallic thallium (Bloembergen and Rowland, 1953).

long to cause appreciable broadening of the resonance, thus ruling out (c) also. In fact, within experimental error, the experimental second moment of the resonance spectrum for these metals is in agreement with that derived theoretically using Van Vleck's formula (6.13), and the known lattice parameter of the metals (Gutowsky and McGarvey, 1952).

However, except at low temperatures, source (c) is usually of importance also. The conduction electron relaxation mechanism usually causes T_1 to be as short as 10^{-3} sec. at room temperature, thus endowing the resonance line with a width of several tenths of a gauss. This is in marked contrast with rigid non-metallic solids of the kind discussed in §§6.2 and 6.3, for which T_1 is usually at least 10 sec., thus endowing the line with quite negligible breadth. Gutowsky and McGarvey (1952) found that the spectra for rubidium (^{85}Rb and ^{87}Rb) and caesium (^{133}Cs) at room temperature derive their breadth almost entirely from source (c), whilst the second moments of aluminium (^{27}Al) and copper (^{63}Cu and ^{65}Cu) are augmented from this source by 23 and 13% respectively.

Sodium exhibits a narrowing of the resonance line with increase of temperature as shown in fig. 54 (Gutowsky, 1951). Since for

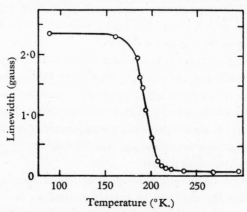

Fig. 54. The resonance line width transition for ^{23}Na in metallic sodium (Gutowsky, 1951). The narrowing is ascribed to self-diffusion. The line width is here taken as the interval between points of maximum and minimum slope of the absorption line.

sodium the line width at 77° K. originates in nuclear magnetic interaction, the reduction of the width almost to zero must be explained by self-diffusion of the sodium ions (cf. self-diffusion in non-metallic solids, discussed in §6.5). This self-diffusion in sodium has been established directly by a radioactive tracer method using the unstable isotope ^{22}Na (Nachtrieb, Catalano and Weil, 1952). Norberg and Slichter (1951) found that the decrease in resonance line width with increase in temperature could be accounted for quantitatively using the diffusion coefficient and its activation energy determined by Nachtrieb *et al.* Self-diffusion is also evidenced by a narrowing of line width in lithium (Gutowsky and McGarvey, 1952), aluminium (Seymour, 1953) and rubidium, and probably occurs in caesium also (McGarvey and Gutowsky, 1953).

Turning now to non-cubic metals, we first consider white tin, which is tetragonal. Since ^{117}Sn and ^{119}Sn both have spin number $\frac{1}{2}$ they possess no quadrupole moment and there is therefore no contribution to the line breadth from source (b). A contribution can, however, come from source (d), the anisotropy of the resonance shift, as we discussed in §7.3.

All four sources must be expected to contribute to the spectrum of beryllium, since its crystal structure has hexagonal symmetry and the spin number of ^{9}Be is $\frac{3}{2}$. While sources (a), (c) and (d) are all likely to make appreciable contributions to the width of the spectrum, its shape is dominated by source (b), the quadrupole interaction. Knight (1953) has found the spectrum to be a well-resolved triplet, which, as we shall see in §8.3, is the expected form for a polycrystalline specimen with quadrupole splitting from nuclei of spin number $\frac{3}{2}$.

Quadrupole broadening may also be exhibited by metals such as copper whose crystal structure is normally cubic, if the metal is in a strained condition. If the crystal lattice is distorted from its cubic shape, the electric field gradient is no longer zero at the nuclear sites and quadrupole broadening results. Thus the ^{63}Cu resonance in cold-worked copper filings has only about half the intensity of that for well-annealed copper (Bloembergen and Rowland, 1953); ^{63}Cu has spin number $\frac{3}{2}$, and the cold work causes the line to be broken up into a triplet, the two satellites of which are too broad to be observed. A similar effect in an ionic salt was discussed in §4.3.1.

Bloembergen and Rowland (1953) have also shown that quadrupole broadening is important in disordered alloys of cubic crystal structure. Thus, for example, in cubic α-brass the environment of individual copper atoms in the solid solution does not have cubic symmetry. The intensity of the ^{63}Cu resonance therefore decreases rapidly due to quadrupole broadening as the concentration of zinc atoms increases; with an atomic proportion of 10% Zn the reduction of intensity is by a factor 0·15. In ordered alloys with cubic structure quadrupole broadening should be absent. The nuclear magnetic resonance signal can thus give information about the state of short-range order and disorder in cubic alloys.

7.5. Liquid metals

Nuclear magnetic resonance has been observed in gallium, rubidium and caesium in the liquid state (Knight, 1949; Gutowsky and McGarvey, 1952; McGarvey and Gutowsky, 1953). As one might expect, a resonance shift is found just as with solid metals. In fact, when rubidium and caesium pass through their melting-points there is only a small discontinuity in the shift, which is associated with the discontinuity in density. The breadth of the resonance line for all three liquid metals is several tenths of a gauss, and is most probably associated with a short spin-lattice relaxation time, although values of T_1 have not been directly measured.

CHAPTER 8

QUADRUPOLE EFFECTS

8.1. Introduction

Nuclei of spin number I greater than $\frac{1}{2}$ can, and usually do, possess electric quadrupole moments. The quadrupole moment is a measure of the departure of the nuclear charge distribution from spherical symmetry. Such nuclei interact with electrostatic potentials of lower than cubic symmetry, and the energy of interaction is therefore a function of the orientation of the nucleus with respect to its electrostatic environment, and hence of the magnetic quantum number m which describes this orientation. If the electrostatic environment is static, as in a crystal, the magnetic sub-levels of energy are therefore perturbed unequally by this interaction, and in consequence the nuclear magnetic resonance line for a crystalline specimen is split into a number of component lines. In a liquid the molecular motion causes the interaction to be time-varying with an average value which is in general much less than for a crystal, as in the case of the magnetic dipolar interaction discussed in Chapter 5. Indeed, qualitatively the effects of quadrupole interaction upon the spectrum and relaxation time of solids and liquids are the same as the effects of the magnetic dipolar interaction which we have already discussed. Quantitatively, however, the quadrupole effects are always much larger, and we will now show this by an order of magnitude argument.

The energy of magnetic interaction between a nucleus of magnetic moment μ with an identical neighbour at a distance r is of the order μ^2/r^3. The energy of electric quadrupolar interaction is of the order of the product of the nuclear electric quadrupole moment†

---

† The scalar eQ, where e is the electronic charge and Q has dimensions (length)^2, is defined by

$$eQ = \int \rho(3z'^2 - r^2)\,dV, \qquad (8.1)$$

where dV is any volume element in the nucleus, r is its distance from the centre of the nucleus, z' is its co-ordinate along the symmetry axis of the nucleus, and ρ is the electric charge density. If the charge distribution is spherically symmetrical, integration of (8.1) shows that Q is zero.

eQ and the electric field gradient. If the near nuclear neighbour is at the centre of a charged ion, the electric field gradient is of order e/r^3, and hence the interaction energy is of order e^2Q/r^3. The ratio of electric quadrupole and magnetic dipole interaction energies in such cases should thus be of order e^2Q/μ^2. Values of this ratio for nuclear species with $I > \frac{1}{2}$ fall mainly in the range 10–10^3; the smallest value appears to be that of ^6Li, which is about 5. If the atom containing the nucleus of interest is bound by bonds having even partial covalent character, the electric field gradient is much greater than our rough estimate. We therefore conclude that if quadrupole interactions are present they predominate over the nuclear magnetic dipolar interactions.

The next section (§8.2) is concerned with quadrupole effects in liquids and gases. In §§8.3 and 8.4 their effects on the spectrum and spin-lattice relaxation of solids are respectively discussed. In a final section (§8.5) we briefly discuss pure quadrupole resonance, which strictly speaking falls outside the province of this monograph, but is closely related to the topics with which this chapter is concerned.

8.2. Quadrupole effects in liquids and gases

It was seen in §5.2 that the interaction between the nuclear magnetic moment and the fluctuating local magnetic field provided the dominant source of spin-lattice relaxation in liquids for nuclei of spin number $\frac{1}{2}$. When the spin number exceeds $\frac{1}{2}$ we therefore have an additional relaxation mechanism provided by the interaction between the nuclear electric quadrupole moment and the fluctuating local electric field gradient. A theory of this relaxation process would follow similar lines to that for the magnetic process. One expects that the formula for the relaxation time T_1 should be of similar form to equation (5.14), but that wherever the magnetic interaction energy μ^2/r_0^3 appears it should be replaced by a quantity of order $eQ(\partial^2 \mathscr{V}/\partial z^2)$, descriptive of the quadrupole interaction energy. ($\mathscr{V}$ is the electrostatic potential, so that $-\partial \mathscr{V}/\partial z$ is the z component of the electric field, and $-\partial^2 \mathscr{V}/\partial z^2$ is the z component of its gradient.) Such a theory is developed by Bloembergen (1948), but we shall not give the details here since a quantitative comparison between theory and experiment is rendered difficult by the lack of knowledge of the electric field gradient (see also Ayant, 1954).

The quadrupole contribution to spin-lattice relaxation is demonstrated by measurements made by Bloembergen, Purcell and Pound (1948) on a mixture of 50% H_2O and 50% D_2O. The relaxation times for the proton and the deuteron resonances were 3·0 and 0·5 sec. respectively. If the relaxation of the deuterons was entirely magnetic as for the protons, then from the considerations of §5.2 we should expect the deuteron relaxation time to be the longer of the two, rather than the shorter, since the magnetic moment of the deuteron is several times smaller than that of the proton. Bloembergen *et al.* found that the observed value of T_1 for the deuteron resonance could be accounted for by assuming a reasonable value for $\partial^2 \mathscr{V} / \partial z^2$. The quadrupole moment of the deuteron is relatively small, and for other nuclei with greater Q the relaxation time T_1 is often very short and endows the resonance line with an appreciable breadth. Thus Pound (1947 *b*) found resonance lines 10 gauss broad for [79]Br and [81]Br in aqueous solutions of lithium bromide and sodium bromide, indicative of a relaxation time of order 3×10^{-5} sec.

As we mentioned in §4.4 the relative widths of relaxation-broadened lines for two isotopes of the same atomic species in the same specimen furnishes a rough value of the ratio of their quadrupole moments. Since in equations (5.14) and (5.19) the expressions for $1/T_1$ contain the square of the magnetic interaction energy μ^2 / r^3 as a factor, the argument at the beginning of this section shows that the corresponding expressions for quadrupole relaxation contain the factor Q^2. Since both nuclear species are subjected to the same electric environment, the resonance line widths are therefore proportional to Q^2.

Quadrupole interactions should contribute to the spin-lattice relaxation process for gaseous specimens also. Bloembergen (1948) has given preliminary consideration to the problem, and concludes that the relaxation time for gaseous deuterium should be less than for gaseous hydrogen at the same pressure.

8.3. Quadrupole effects on the nuclear magnetic resonance spectrum for solids

The foundations of our knowledge of this branch of the subject were laid by the work of Pound (1950), who carried out a series of important experiments with accompanying theoretical analysis.

In §2.1 we showed that the magnetic energy levels for an isolated nucleus of magnetic moment μ and spin number I subjected to a magnetic field $\mathbf{H}_0$ directed along the z axis are given by

$$E_m = - m\mu H_0/I, \tag{8.2}$$

where m is the magnetic quantum number. If the nucleus possesses an electric quadrupole moment $(I > \frac{1}{2})$, it interacts with its electric environment if this has less than cubic symmetry. If the quadrupole interaction is weak compared with the magnetic interaction with $\mathbf{H}_0$, it may be treated as a small perturbation. Using first-order perturbation theory, Pound (1950) finds that the energy levels now become

$$E_m = - \frac{m\mu H_0}{I} + \frac{eQ}{4I(2I-1)} \left[3m^2 - I(I+1) \right] \frac{\partial^2 \mathscr{V}}{\partial z^2}, \tag{8.3}$$

the second term representing the displacement of the energy levels of (8.2) by the quadrupole interaction. The $2I$ intervals between successive energy levels are now no longer all equal, but instead are all different. If, therefore, we have a single crystal containing nuclei which are all disposed at equivalent lattice sites, the resonance spectrum is split into $2I$ component lines, symmetrically placed on either side of the position the line would take in absence of quadrupole interaction. We note that if m is a half-integer, the transition $m = \frac{1}{2}$ to $m = -\frac{1}{2}$ gives a component line in this central position. If there are non-equivalent nuclear sites in the crystal lattice there will be a set of $2I$ component lines for each type of site. The resonance of ^{23}Na, for which $I = \frac{3}{2}$, in a single crystal of sodium nitrate is shown in fig. 55 (Pound, 1950); the three component lines are well resolved. For this orientation of the crystal with respect to $\mathbf{H}_0$ the separation between the central line and either satellite is 84 kc./s. on a frequency scale, or 75 gauss in terms of magnetic field. The splitting is thus an order of magnitude greater than that due to magnetic dipolar interaction between the sodium nuclei. This magnetic interaction is in fact largely respon-

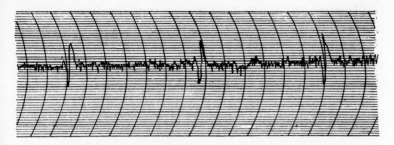

Fig. 55. The nuclear magnetic resonance absorption derivative spectrum of ^{23}Na in a single crystal of sodium nitrate with the applied magnetic field $\mathbf{H_0}$ perpendicular to the trigonal crystal axis (Pound, 1950). The frequency of the centre line is 7·18 Mc./s. ($H_0 = 6379$ gauss). The separation between the central line and either satellite is 84 kc./s., equivalent to 75 gauss. A broad weak resonance from the ^{63}Cu nuclei in the wire of the specimen coil is to be seen on the right of the central line.

sible for the 2–3 gauss width of the component lines, and may be neglected when one is solely interested in analysing the quadrupole effects.

The splitting is proportional to $\partial^2 \mathscr{V} / \partial z^2$ which is a function of direction relative to the crystal axes. Consequently the spectrum for powdered specimens, containing randomly oriented crystallites, is spread out over a wide range of frequency (or field strength). For odd values of I, the central component line ($m = \frac{1}{2} \rightarrow m = -\frac{1}{2}$) remains fixed according to (8.3). However, if the quadrupole interaction is moderately strong the first-order perturbation of (8.3) is not a sufficiently accurate description, and in the second and higher even orders this component line also has angular dependence and is spread out for a powdered specimen. As a consequence, resonances in powders, if moderate quadrupole coupling exists, are almost unobservably weak. Fig. 56 shows the ^7Li resonance in powdered Li_2CO_3 in which the quadrupole interaction is relatively weak.

If each nucleus is situated in an electric environment having axial symmetry, the field gradient component is given by (Pound, 1950)

$$\frac{\partial^2 \mathscr{V}}{\partial z^2} = \tfrac{1}{2} eq (3 \cos^2 \theta - 1), \tag{8.4}$$

where θ is the angle between the symmetry axis and $\mathbf{H_0}$ (along the

z direction), and eq is a scalar descriptive of the electric environment defined as

$$eq = \int \rho(3 \cos^2 \theta - 1) r^{-3} \, dV, \qquad (8.5)$$

where the integral is taken over all charges outside the nucleus, $\mathbf{r}$ is the vector joining the nucleus to the volume element dV, and θ the

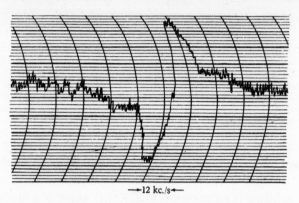

$\longrightarrow$ 12 kc./s $\longleftarrow$

Fig. 56. The resonance absorption derivative of ^{7}Li in polycrystalline lithium carbonate (Pound, 1950).

angle between $\mathbf{r}$ and the symmetry axis. Using (8.3) and (8.4) we find that the component lines should fall at frequencies

$$\nu_{m \to m-1} = \nu_0 + \frac{3e^2 Qq(2m-1)}{8I(2I-1)h} (3 \cos^2 \theta - 1), \qquad (8.6)$$

where ν_0 is the unperturbed resonance frequency $\mu H_0/Ih$ given by (2.1). Pound (1950) tested this relation for the three component lines of ^{23}Na ($I = \frac{3}{2}$) in a single crystal of sodium nitrate. The sodium ions lie on a threefold symmetry axis of the crystal. The crystal was rotated about an axis perpendicular to $\mathbf{H_0}$, and perpendicular to the triad axis of the crystal. The spectrum when the angle θ between $\mathbf{H_0}$ and the triad axis was $90°$ is shown in fig. 55. The frequency interval between the central line and the symmetrically placed satellites is plotted as a function of θ in fig. 57. The experimental points are a good fit to the smooth curve given by $83\cdot5 \, (3 \cos^2 \theta - 1)$ kc./s., and hence one concludes from (8.6), with $I = \frac{3}{2}$, that $\frac{1}{4}e^2 Qq/h$ is $83\cdot5$ kc./s. Unfortunately, it is not possible at present to calculate the value of eq for solids, using (8.5), since the local charge distribution is insufficiently well known; one cannot therefore deduce the value of the nuclear quadrupole moment from these experiments.

Pound (1950) found a much stronger quadrupole interaction for the ^{27}Al nuclei ($I=\frac{5}{2}$) in a single crystal of Al_2O_3. The ^{27}Al nuclei

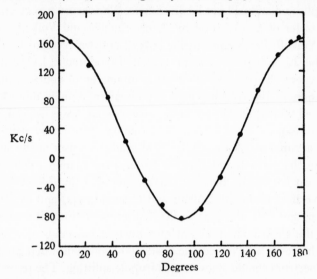

Fig. 57. The frequency interval between the central line and the symmetrically placed satellites for the ^{23}Na spectra in sodium nitrate, as a function of the angle θ between H_0 and the crystal axis (Pound, 1950).

are situated on a hexagonal symmetry axis. The spectrum showed the anticipated five component lines, with a maximum splitting from centre to far satellite of 720 kc./s. The angular dependence of the splitting showed appreciable divergences from the $(3 \cos^2 \theta - 1)$ variation indicated by first-order perturbation theory from equations (8.3) and (8.4). Moreover, the central line did not remain fixed, but was a function of θ, departing by as much as 40 kc./s. from ν_0. The spectrum was measured in a relatively low applied field of about 2700 gauss, for which ν_0 is about 3 Mc./s. The splitting of ±720 kc./s. is thus about $\pm25\%$ of the central frequency, and thus constitutes more than a small perturbation. Pound (1950) therefore carried the perturbation calculation to the third order, and obtained excellent agreement between observed and calculated frequencies for the five component lines at all angles. The theory has also been discussed by Carr and Kikuchi (1950) and has been notably extended by Bersohn (1952).

The work of Pound (1950) was confined to uniaxial crystals to which equation (8.4) could be applied. The theory for crystals containing nuclei at sites devoid of axial symmetry is much more complicated and will not be given here. The theory was given to a first order of perturbation by Petch, Smellie and Volkoff (1951) and Volkoff, Petch and Smellie (1952), and was verified by them for the ^{7}Li resonance in a single crystal of spodumene, $LiAl(SiO_3)_2$. The spectrum still consists of $2I$ components of course, but the variation of the splitting with crystal orientation is more complicated than (8.4). The coupling constant e^2Qq/h was derived, and, in addition, the orientation of the principal axes and the degree of axial asymmetry of the electric field gradient tensor at the site of the nuclei were found. Higher order perturbation calculations were carried out by Bersohn (1952), extended by Volkoff (1953), and verified by Petch, Volkoff and Cranna (1952) and by Petch, Cranna and Volkoff (1953) for the ^{27}Al resonance in spodumene.

If the electric environment of a nucleus has cubic symmetry, then however large an electric quadrupole moment it may possess, the spectrum should show no quadrupole splitting. The resonance of ^{127}I (spin number $\frac{5}{2}$) in a single crystal of potassium iodide verifies this (Pound, 1950); the resonance line width of 0·7 gauss was determined by the magnetic dipolar interaction between the nuclei. If the ^{127}I nucleus possesses an electric sedicipole moment, which as a nucleus of spin number $\frac{5}{2}$ it may, an electric interaction is possible even in a cubic environment, involving the fourth derivative of the potential. Knowledge of the strength of $\partial^4 \mathscr{V}/\partial z^4$ at the nuclear sites would allow an upper limit to be set on the magnitude of the sedicipole moment, since Pound (1950) has found that the splitting due to this interaction must be less than about 200 c./s.

If a single crystal of cubic symmetry is work-hardened, the electric environment of the nuclei is no longer of cubic symmetry in the resulting strained crystal. Moreover, the value of $\partial^2 \mathscr{V}/\partial z^2$ is in general different for each nucleus, and the spectrum is spread over a wide frequency range. If I is an odd half-integer, each nucleus continues to give a central line at frequency ν_0, and this may well be the only part of the spectrum observed (Watkins and

Pound, 1953; Pound, 1953; Bloembergen and Rowland, 1953). Even the remaining central line is broadened by the second-order perturbation shifts if the strain is great.

8.4. Quadrupole relaxation effects in solids

In §6.6 we discussed spin-lattice relaxation mechanisms for nuclei which have no electric quadrupole moment, and discovered that the mechanism based on the paramagnetic relaxation theory of Waller (1932) was too weak. In Waller's theory the thermal vibrations of the lattice provided a means of energy exchange between the spin system and the lattice. The theory led to values of T_1 of at least 10^4 sec. at room temperature, and to very much greater values at lower temperatures.

For nuclei with $I > \frac{1}{2}$, a mechanism analogous to Waller's, but using the electric quadrupole interaction of the nuclei with their environment rather than the magnetic dipolar interaction, is more encouraging. The component at frequency ν_0 of the Fourier spectra of the electric field gradient enables the lattice to induce transitions changing the nuclear magnetic quantum number m by ± 1, thus exchanging energy with the spin system. The component at frequency $2\nu_0$ is also able to excite transitions, since the quadrupole interaction operator has matrix elements corresponding to changes of m by ± 2 as well as ± 1.

We saw in §8.1 that the ratio of the electric quadrupole interaction energy of a nucleus with its environment to the corresponding magnetic dipolar interaction energy is of the order of $e^2 Q / \mu^2$, values of which fall mainly in the range 10–10^3. In §5.2 and Appendix 3 we saw that the relaxation time T_1 is inversely proportional to the square of the interaction energy. We therefore expect the relaxation time brought about by lattice vibrations to be shorter by a factor of order $(e^2 Q / \mu^2)^2$ when quadrupole interactions are present. This brings the theoretical values of T_1 at room temperature down to the region of one second, and comparable with the values experimentally observed. It is to be noticed that this mechanism should exist even in cubic crystals; as the lattice vibrates the electrostatic potentials at the nuclear sites are disturbed from their time-averaged cubic symmetry, leading to a time-varying quadrupole interaction.

In a series of elegant experiments Pound (1950) was able to show beyond doubt that the electric quadrupole interaction provided the operative relaxation mechanism in a number of crystals. First there was the qualitative evidence that T_1 for the ^{127}I resonance in KI was much shorter than for the ^{23}Na resonance in NaNO$_3$, which in turn was much shorter than for the ^{7}Li resonance in LiNO$_3$. This accords with our knowledge that the quadrupole moment for ^{127}I is considerably greater than that for ^{7}Li, while that for ^{23}Na probably lies between.

Further evidence came from the study of the cubic crystal NaBr. The spin numbers of ^{23}Na, ^{79}Br and ^{81}Br are all $\frac{3}{2}$, and the three magnetic moments are almost equal (within 8%). Because the magnetic moments are so nearly equal, any relaxation process depending on magnetic interaction should result in nearly equal relaxation times. In fact, however, Pound found that the relaxation time for ^{23}Na was about twice as long as for ^{81}Br. Thus a magnetic relaxation mechanism seems to be excluded and an electric quadrupole mechanism seems likely.

The most convincing experiment consisted in observing the signal from one component of a quadrupole-split spectrum, whilst another component was saturated by a radiofrequency field of large

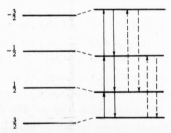

Fig. 58. Nuclear magnetic energy levels perturbed by electric quadrupole interaction for $I = \frac{3}{2}$. The full vertical lines indicate transitions which may be brought about by magnetic interaction between the nuclear spin system and the lattice; the broken vertical lines indicate additional transitions which are possible when the spin-lattice interaction originates in electric coupling of the lattice to the nuclear electric quadrupole moments.

amplitude. It is necessary to carry out an analysis of this situation in order to see its relevance. The experiment was carried out using the ^{23}Na resonance from a single crystal of NaNO$_3$; the spin number is $\frac{3}{2}$, and we therefore consider this value of I in the following discussion. The four magnetic energy levels of the ^{23}Na nucleus, perturbed by the quadrupole interaction, are shown in fig. 58; the transitions corresponding to the three components of the spectrum (see fig. 55) are indicated by full vertical lines.

First consider the case of a magnetic spin-lattice relaxation process. Let us apply a radiofrequency field at the frequency corresponding to transitions between the levels $m = -\frac{3}{2}$ and $m = -\frac{1}{2}$. As in §2.4 we let $P_{m \to m'}$ represent the probability of transitions induced by the applied signal from state m to state m'. As in §2.2 we let $W_{m \to m'}$ represent the probability of transitions brought about by the relaxation process. If N_m is the number of nuclei per cm.^{3} in state m, then

$$\left.\begin{aligned}
\frac{dN_{-\frac{3}{2}}}{dt} &= -(W_{-\frac{3}{2} \to -\frac{1}{2}} + P_{-\frac{3}{2} \to -\frac{1}{2}})N_{-\frac{3}{2}} \\
&\qquad + (W_{-\frac{1}{2} \to -\frac{3}{2}} + P_{-\frac{1}{2} \to -\frac{3}{2}})N_{-\frac{1}{2}}, \\
\frac{dN_{-\frac{1}{2}}}{dt} &= (W_{-\frac{3}{2} \to -\frac{1}{2}} + P_{-\frac{3}{2} \to -\frac{1}{2}})N_{-\frac{3}{2}} \\
&\qquad - (W_{-\frac{1}{2} \to -\frac{3}{2}} + W_{-\frac{1}{2} \to \frac{1}{2}} + P_{-\frac{1}{2} \to -\frac{3}{2}})N_{-\frac{1}{2}} + W_{\frac{1}{2} \to -\frac{1}{2}}N_{\frac{1}{2}}, \\
\frac{dN_{\frac{1}{2}}}{dt} &= W_{-\frac{1}{2} \to \frac{1}{2}}N_{-\frac{1}{2}} - (W_{\frac{1}{2} \to -\frac{1}{2}} + W_{\frac{1}{2} \to \frac{3}{2}})N_{\frac{1}{2}} + W_{\frac{3}{2} \to \frac{1}{2}}N_{\frac{3}{2}}, \\
\frac{dN_{\frac{3}{2}}}{dt} &= W_{\frac{1}{2} \to \frac{3}{2}}N_{\frac{1}{2}} - W_{\frac{3}{2} \to \frac{1}{2}}N_{\frac{3}{2}}.
\end{aligned}\right\} \quad (8.7)$$

We see from equation (2.21) that the probability for magnetically induced transitions is proportional to $(I+m)(I-m+1)$. Consequently we may write for the upward transitions

$$W_{\frac{3}{2} \to \frac{1}{2}} = W_{-\frac{1}{2} \to -\frac{3}{2}} = 3w, \quad W_{\frac{1}{2} \to -\frac{1}{2}} = 4w; \quad (8.8)$$

and also

$$P_{-\frac{3}{2} \to -\frac{1}{2}} = P_{-\frac{1}{2} \to -\frac{3}{2}} = 3pw. \quad (8.9)$$

The probabilities of downward transitions brought about by thermal processes are given by those for the corresponding upward transitions multiplied by the factor $\exp(h\nu_0/kT)$, as we explained in §2.2. Thus

$$\left.\begin{aligned}
W_{\frac{1}{2} \to \frac{3}{2}} &= W_{-\frac{3}{2} \to -\frac{1}{2}} = 3w \exp(h\nu_0/kT) = 3w(1 + \Delta), \\
W_{\frac{1}{2} \to -\frac{1}{2}} &= 4w(1 + \Delta),
\end{aligned}\right\} \quad (8.10)$$

where

$$\Delta = h\nu_0/kT. \quad (8.11)$$

Substituting from (8.8), (8.9) and (8.10) into equations (8.7) we get

$$
\left.
\begin{aligned}
\frac{dN_{-\frac{3}{2}}}{dt} &= w\{-3(1+\varDelta+p)N_{-\frac{3}{2}}+3(1+p)N_{-\frac{1}{2}}\}, \\
\frac{dN_{-\frac{1}{2}}}{dt} &= w\{3(1+\varDelta+p)N_{-\frac{3}{2}}-(7+4\varDelta+3p)N_{-\frac{1}{2}}+4N_{\frac{1}{2}}\}, \\
\frac{dN_{\frac{1}{2}}}{dt} &= w\{4(1+\varDelta)N_{-\frac{1}{2}}-(7+3\varDelta)N_{\frac{1}{2}}+3N_{\frac{3}{2}}\}, \\
\frac{dN_{\frac{3}{2}}}{dt} &= w\{3(1+\varDelta)N_{\frac{1}{2}}-3N_{\frac{3}{2}}\}.
\end{aligned}
\right\}
\tag{8.12}
$$

If a steady state is maintained, all terms on the left-hand side of (8.12) are zero and we obtain

$$
N_{\frac{3}{2}}=(1+\varDelta)N_{\frac{1}{2}}, \quad N_{\frac{1}{2}}=(1+\varDelta)N_{-\frac{1}{2}}, \quad N_{-\frac{1}{2}}=[1+\varDelta/(1+p)]N_{-\frac{3}{2}}.
\tag{8.13}
$$

With the restriction that

$$
N_{\frac{3}{2}}+N_{\frac{1}{2}}+N_{-\frac{1}{2}}+N_{-\frac{3}{2}}=N,
$$

where N is the total number of nuclei per cm.[3], we see that the differences in population $(N_{\frac{3}{2}}-N_{\frac{1}{2}})$ and $(N_{\frac{1}{2}}-N_{-\frac{1}{2}})$ are independent of p to a first order in $\varDelta$, being given by $\frac{1}{4}N\varDelta$, whereas the third difference is given by

$$
N_{-\frac{1}{2}}-N_{-\frac{3}{2}}=\frac{N\varDelta}{4(1+p)}=\frac{n_0}{1+p},
\tag{8.14}
$$

where n_0 is the equilibrium population difference for all three transitions when $p=0$. Thus the application of a large signal $(p\gg 1)$ at the frequency of the $m=-\frac{3}{2}$ to $m=-\frac{1}{2}$ transition has no effect on the strength of the signals from the other two transitions measured at low signal strength, though its own transition is saturated. If we repeat the calculation with the strong signal applied to the $m=-\frac{1}{2}$ to $m=\frac{1}{2}$ transition we find that to a first order in $\varDelta$, only the difference $(N_{-\frac{1}{2}}-N_{\frac{1}{2}})$ is modified. Thus we conclude that provided $h\nu_0 \ll kT$, the application of a large signal at one transition frequency does not affect the absorption of a small signal at the other two component lines, if the relaxation process is magnetic in origin.

A different result is obtained if electric quadrupole interaction causes the relaxation. The difference arises because transitions involving changes of m by ± 2 are now also possible (shown as dashed lines in fig. 58). We may now carry out an exactly similar calculation to the previous one and find this time that

$$
\begin{aligned}
N_{-\frac{1}{2}} - N_{-\frac{3}{2}} &= n_0 \left[\frac{1}{1 + \frac{9}{4}p} \right], \\
N_{\frac{1}{2}} - N_{-\frac{1}{2}} &= n_0 \left[\frac{1 + \frac{15}{4}p}{1 + \frac{9}{4}p} \right], \\
N_{\frac{3}{2}} - N_{\frac{1}{2}} &= n_0 \left[\frac{1 + \frac{3}{2}p}{1 + \frac{9}{4}p} \right],
\end{aligned}
\tag{8.15}
$$

when a strong signal is applied whose frequency corresponds to transitions between the state $m = -\frac{3}{2}$ and $m = -\frac{1}{2}$. Thus we see that in this case *all* the differences in population are disturbed by the applied signal. The difference $(N_{\frac{1}{2}} - N_{-\frac{1}{2}})$ increases from n_0 (when $p = 0$) to $\frac{5}{3}n_0$ ($p \gg 1$), and the strength of this component line, measured with small signal strength, should increase by a factor $\frac{5}{3}$. Similarly, the strength of the line corresponding to $m = \frac{3}{2}$ to $m = \frac{1}{2}$ should decrease by a factor $\frac{2}{3}$. Pound observed that the intensity of these two component lines did in fact change by the predicted factors $\frac{5}{3}$ and $\frac{2}{3}$ respectively. This agreement thus furnishes very strong evidence that the electric quadrupole interaction provides the dominant spin-lattice relaxation process in $NaNO_3$.

8.5. Pure quadrupole resonance

We noticed in §8.3 that for Al_2O_3 the perturbation of the nuclear magnetic energy levels by the nuclear electric quadrupole inter-action was an appreciable fraction of the separation between these levels produced by a magnetic field of several kilogauss. In many solids the electric quadrupole interactions are still greater, and far larger than the splittings produced by reasonable magnetic fields. This is particularly true of molecular solids where the electric field gradient is intramolecular in origin and has the same order of magnitude as in free molecules. For covalently bonded molecules, typical values of e^2Qq/h for ^{35}Cl, ^{79}Br and ^{127}I are 80, 500 and 2000 Mc./s. respectively. It is clear that in such cases the quadru-pole interaction is primarily responsible for resolving the degen-eracy, and that any magnetic field which may be present must be treated as a perturbation of the quadrupole interaction. In such solids there is thus a large splitting even in absence of a magnetic

P

field, and the possibility therefore arises of obtaining a resonant exchange of radiofrequency energy with the solid even in absence of an applied magnetic field.

The first successful experiment was carried out by Dehmelt and Krüger (1950), who found resonances for ^{35}Cl and ^{37}Cl (both of spin number $\frac{3}{2}$) in solid *trans*-1,2-dichloroethylene at frequencies of 35·40 and 27·96 Mc./s. respectively. The phenomenon might still properly be regarded as nuclear *magnetic* resonance in that the coupling between the nuclei and the electromagnetic radiation which brings about the nuclear transitions is magnetic, as we shall see presently. Nevertheless, the energy levels are split electrically and not magnetically, and for this reason the phenomenon is usually called *nuclear quadrupole resonance* or *pure quadrupole resonance* in order to emphasize the absence of any applied magnetic field. Pure quadrupole resonance has now become a separate field of study, and we shall give only a brief discussion here; for a more detailed discussion the reader is referred to the excellent review by Dehmelt (1954).

Let us consider first from a classical point of view a nucleus possessing an electric quadrupole moment situated in an electric environment having axial symmetry, the axis of symmetry being taken along the z axis. Let the axis of spin of the nucleus, and therefore also the magnetic moment vector, be inclined to the z

Fig. 59. Diagram illustrating the precession of a nuclear quadrupole about the axis of symmetry of its electric environment.

axis with angle θ (see fig. 59). Then, as shown by Dehmelt (1954) the classical quadrupole interaction energy is

$$E = \tfrac{1}{8}e^2 Q q (3 \cos^2 \theta - 1). \tag{8.16}$$

Since E is a function of θ, a couple $dE/d\theta$ must act upon the nucleus tending to turn the nuclear axis towards or away from the z axis, according to the sign of Qq. The spinning nucleus responds to this couple by precessing about the z axis. In consequence, there are components of the magnetic dipole moment and of the electric quadrupole moment rotating about the z axis. If, therefore, electromagnetic radiation of the precessional frequency is applied with the magnetic or electric vector rotating in the same sense about the z axis, the angle θ will change, as we saw in §2.1; in practice, however, only the rotating magnetic dipole experiences appreciable coupling with the radiation since that of the electric quadrupole is very much smaller.

The quantum-mechanical expression for the energy levels, equivalent to the classical formula (8.16), is

$$E_m = \tfrac{1}{4}e^2 Qq \, \frac{[3m^2 - I(I+1)]}{I(2I-1)}. \tag{8.17}$$

This leads to the energy levels shown in fig. 60 for integral and

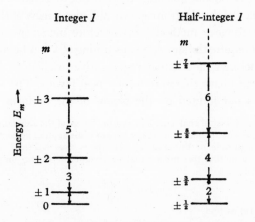

Fig. 60. Energy levels arising from the interaction of a nuclear electric quadrupole with an axially symmetric electric field. The unit of energy is $3e^2Qq/4I(2I-1)$ (Dehmelt, 1954).

half-integral values of I. It will be noticed that unlike the nuclear Zeeman levels produced by a magnetic field, the separations are all unequal, and transitions between successive levels require different radiofrequencies.† For $I = \frac{5}{2}$ it is seen that two resonance lines should be observed whose frequencies are in the ratio $2 : 1$. Both frequencies were observed for ^{127}I (spin number $\frac{5}{2}$) in a number of iodine compounds (Dehmelt, 1951). The ratio of the frequencies differed by a small percentage from 2, and the departure from 2 was shown to be a measure of the deviation of the electrostatic potential from axial symmetry. Pure quadrupole resonance may be detected experimentally using the same techniques as for nuclear magnetic resonance absorption, discussed in Chapter 3. The specimen, which need not be a single crystal, is placed inside a radiofrequency coil which provides electromagnetic radiation when energized by radiofrequency current. Any absorption of radiofrequency energy by the specimen is demonstrated by suitable measurement of the accompanying diminution of the $\mathcal{Q}$ of the coil. Whereas the frequency of resonance is determined in nuclear magnetic resonance by the strength of the applied magnetic field, which one is free to determine, in pure quadrupole resonance the frequency is fixed by the internal electric field gradient within the crystal. The resonance line must therefore be detected by traversing the frequency of the electromagnetic radiation; many of the nuclear magnetic resonance methods are therefore inconvenient because they would require the simultaneous tuning of a number of circuits. The regenerative and super-regenerative oscillator techniques (§§3.4 and 3.5) have therefore found most favour. The size of the specimen is not limited by the extent of the uniform field of a

† This has its classical analogue. The rate of change of the angular momentum $\mathbf{\Omega}$ of the precessing nucleus is $\mathbf{\omega} \wedge \mathbf{\Omega}$, where $\mathbf{\omega}$ is the angular velocity of precession. The magnitude of this angular velocity is thus $\omega\Omega \sin \theta$, and by the classical law of motion this must equal the magnitude of the applied couple. Hence we have

$$\omega\Omega \sin \theta = \frac{dE}{d\theta}, \tag{8.18}$$

and using (8.16) we find

$$\omega = 3e^2 Qq \cos \theta / 4\Omega.$$

Thus the frequency of precession is a function of θ, and decreases with increasing θ, becoming zero for $\theta = \frac{1}{2}\pi$, and increases again with opposite sense for $\theta > \frac{1}{2}\pi$. In the magnetic case the energy of interaction between the magnetic moment μ and the applied field $\mathbf{H}_0$ is $-\mu H_0 \cos \theta$. If this is substituted for E in (8.18) we get

$$\omega = \frac{\mu}{\Omega} H_0,$$

which is Larmor's theorem, showing that ω in this case is independent of θ.

magnet, but the use of a larger specimen does require more radio-frequency power. Since at room temperature the spin-lattice relaxation time T_1 is usually short, optimum conditions for the detection of a weak resonance line call for still more radiofrequency power. The super-regenerative oscillator is therefore valuable since large radiofrequency power can be developed without undue increase in noise power.

The breadth of the observed quadrupole resonance lines ranges from 10^{-2} to 10^{-5} of the resonance frequency. Two of the sources of line-broadening are common to both nuclear magnetic and nuclear quadrupole resonances; they are (a) magnetic dipolar inter-action between neighbouring nuclei, and (b) broadening consequent upon a short relaxation time T_1. A third source arises from im-perfections in the crystal lattice which cause the field gradient to vary somewhat from molecule to molecule. As crystals are cooled elastic strains within the crystal are accentuated, and this can cause the breadth to increase with decrease of temperature. A fourth source of broadening is caused by torsional vibrations of the mole-cules which modulates the electric field gradient. In addition to broadening the line, the torsional oscillations reduce the mean value of the electric field gradient and thus reduce the resonance frequency. Bayer (1951) proposes on this basis to account for the small decrease in resonance frequency with increase in temp-erature, since the amplitude of the torsional vibrations should in-crease with temperature. Bayer also suggests that the torsional oscillations provide an effective mechanism for spin-lattice relaxation.

The information which can be obtained from nuclear quadrupole resonance measurements will now be mentioned. It will be seen from fig. 60 that for the case of axial symmetry the ratio of the resonance frequencies for two isotopes having the same spin number, for example, ^{35}Cl and ^{37}Cl, in the same crystal gives directly the ratio $^{35}Q/^{37}Q$ of their quadrupole moments. This pro-cedure assumes that the electric field is the same for both isotopes, an assumption which appears to be true to about one part in 10^4. If the spin numbers are different for the two isotopes, then fig. 60 shows that the ratio of the resonance frequencies is equal to the ratio of the quadrupole moments multiplied by a simple and exact numerical factor; for example, if the spin numbers were $\frac{3}{2}$ and 1 the factor would be 2. If the electric field departs appreciably from axial symmetry the argument is unaltered if the isotopes have the same spin number, but a correction must be made if they have different spin numbers. Ratios have been measured for the following isotopic pairs (see Dehmelt, 1954): $^{10,11}B$, $^{35,37}Cl$, $^{63,65}Cu$, $^{79,81}Br$, $^{121,123}Sb$, $^{127,129}I$. The accuracy of the ratios is usually a few parts in 10^4.

In order to obtain absolute values of Q it is necessary to be able to calculate the electric field gradient parameter q. This is possible for simple free molecules, and in some cases aggregation of the molecules in a solid has little influence on q; the intermolecular forces must in such cases be very weak in comparison with the molecular covalent bond, and it is desirable that the molecules be held together solely by the weak van der Waals forces.

By measuring the quadrupole resonance frequency for the same nucleus in different molecules, relative values of the field gradient eq in these molecules are deduced. These data enable information concerning the electronic structure of the molecules to be obtained.

The study of the Zeeman splitting of the resonance line for a single crystal by small magnetic fields yields further structural information. This situation has been studied theoretically by Krüger (1951), Bersohn (1952) and Lamarche and Volkoff (1953). By measuring the splitting for various orientations of the crystal in the magnetic field the departure of the electric field from axial symmetry may be deduced and also the orientation of the principal axes of the electric field gradient tensor relative to the crystal axes.

The large electric quadrupole coupling found in crystals such as iodine, in which for ^{127}I the value of e^2Qq/h is 2153 Mc./s. (Dehmelt, 1951), led Pound (1949) to suggest the use of such materials to obtain spatial alignment of nuclei. In suitable materials the nuclei would occupy predominantly the level $m = \pm I$ if they could be brought into thermal equilibrium at temperatures such that $kT < h\nu$, where ν is a quadrupole resonance frequency. Such temperatures may be obtained by the adiabatic demagnetization technique.

SOLUTIONS OF THE BLOCH EQUATIONS (2.49)

The Bloch equations are (from 2.49):

$$\dot{\mathcal{M}}_x = \gamma(\mathcal{M}_y H_0 + \mathcal{M}_z H_1 \sin \omega t) - \mathcal{M}_x/T_2, \tag{1}$$

$$\dot{\mathcal{M}}_y = \gamma(\mathcal{M}_z H_1 \cos \omega t - \mathcal{M}_x H_0) - \mathcal{M}_y/T_2, \tag{2}$$

$$\dot{\mathcal{M}}_z = -\gamma(\mathcal{M}_x H_1 \sin \omega t + \mathcal{M}_y H_1 \cos \omega t) + (\mathcal{M}_0 - \mathcal{M}_z)/T_1. \tag{3}$$

It is convenient to introduce two new variables u and v, defined by

$$u = \mathcal{M}_x \cos \omega t - \mathcal{M}_y \sin \omega t, \tag{4}$$

$$v = -(\mathcal{M}_x \sin \omega t + \mathcal{M}_y \cos \omega t), \tag{5}$$

which implies that

$$\mathcal{M}_x = u \cos \omega t - v \sin \omega t, \tag{6}$$

$$\mathcal{M}_y = -(u \sin \omega t + v \cos \omega t). \tag{7}$$

Substituting for $\mathcal{M}_x$ and $\mathcal{M}_y$ in (1), (2) and (3) we have

$$\dot{u} + u/T_2 + (\Delta\omega)v = 0, \tag{8}$$

$$\dot{v} + v/T_2 - (\Delta\omega)u = -\gamma H_1 \mathcal{M}_z, \tag{9}$$

$$\dot{\mathcal{M}}_z + \mathcal{M}_z/T_1 - \gamma H_1 v = \mathcal{M}_0/T_1, \tag{10}$$

where

$$\Delta\omega = \gamma H_0 - \omega = \omega_0 - \omega. \tag{11}$$

We first seek a steady-state solution of the Bloch equations (called 'slow passage' by Bloch). In this case $\dot{u} = \dot{v} = \dot{\mathcal{M}}_z = 0$. Eliminating u from (8) and (9) we get

$$v = -\gamma H_1 \mathcal{M}_z T_2 [1 + (\Delta\omega)^2 T_2^2]^{-1}. \tag{12}$$

Substituting for v from (12) in (10) we get

$$\mathcal{M}_z = \mathcal{M}_0 \frac{1 + (\Delta\omega)^2 T_2^2}{1 + (\Delta\omega)^2 T_2^2 + \gamma^2 H_1^2 T_1 T_2}. \tag{13}$$

Using (12) we find

$$v = \frac{-\gamma H_1 \mathcal{M}_0 T_2}{1 + (\Delta\omega)^2 T_2^2 + \gamma^2 H_1^2 T_1 T_2}, \tag{14}$$

and this result taken with (8) gives

$$u = -(\Delta\omega)v T_2 = \frac{\gamma H_1 (\Delta\omega) T_2^2 \mathcal{M}_0}{1 + (\Delta\omega)^2 T_2^2 + \gamma^2 H_1^2 T_1 T_2}. \tag{15}$$

Using (6), (7) and (11), and remembering that $\mathcal{M}_0 = \chi_0 H_0$, we find

$$\mathcal{M}_x = \chi_0 \omega_0 H_1 T_2 \left[\frac{(\omega_0 - \omega)T_2 \cos \omega t + \sin \omega t}{1 + (\omega_0 - \omega)^2 T_2^2 + \gamma^2 H_1^2 T_1 T_2} \right], \qquad (16)$$

and

$$\mathcal{M}_y = \chi_0 \omega_0 H_1 T_2 \left[\frac{-(\omega_0 - \omega)T_2 \sin \omega t + \cos \omega t}{1 + (\omega_0 - \omega)^2 T_2^2 + \gamma^2 H_1^2 T_1 T_2} \right]. \qquad (17)$$

Equations (13), (16) and (17) are used in §2.7.

We now consider the solution of equations (1), (2) and (3) given by Bloch (1946) for what he has termed 'rapid passage', in which one passes through resonance in a time short compared with T_1 or T_2, though still long compared with $(\gamma H_1)^{-1}$. Under these conditions we may expect the solution still to correspond to a precessing resultant magnetic moment, though the magnitude of the moment, and its orientation with respect to the magnetic field, may vary with time. We therefore seek a solution of the form

$$\mathcal{M}_x = \mathcal{M}_1(t) \cos \omega t, \quad \mathcal{M}_y = \mathcal{M}_1(t) \sin \omega t, \qquad (18)$$

the resultant magnetic moment $\mathcal{M}(t)$ being given by

$$\mathcal{M}^2 = \mathcal{M}_x^2 + \mathcal{M}_y^2 + \mathcal{M}_z^2 = \mathcal{M}_1^2 + \mathcal{M}_z^2. \qquad (19)$$

Substituting (18) in (1) and (2), we find these equations are satisfied provided

$$\dot{\mathcal{M}}_1 + \mathcal{M}_1/T_2 = 0, \qquad (20)$$

and

$$\mathcal{M}_1(H_0 - \omega/\gamma) = \mathcal{M}_z H_1. \qquad (21)$$

We define a dimensionless parameter δ by the ratio

$$\delta = (H_0 - \omega/\gamma)/H_1 = (\omega_0 - \omega)/\gamma H_1, \qquad (22)$$

$$= \frac{\mathcal{M}_z}{\mathcal{M}_1}, \qquad (23)$$

using (21). Thus from (19) and (23) we have

$$\mathcal{M}_1 = \frac{\mathcal{M}}{(1 + \delta^2)^{\frac{1}{2}}} \quad \text{and} \quad \mathcal{M}_z = \frac{\mathcal{M}\delta}{(1 + \delta^2)^{\frac{1}{2}}}. \qquad (24)$$

The solution is therefore

$$\mathcal{M}_x = \frac{\mathcal{M}}{(1 + \delta^2)^{\frac{1}{2}}} \cos \omega t, \quad \mathcal{M}_y = \frac{\mathcal{M}}{(1 + \delta^2)^{\frac{1}{2}}} \sin \omega t, \quad \mathcal{M}_z = \frac{\mathcal{M}\delta}{(1 + \delta^2)^{\frac{1}{2}}}, \quad (25)$$

where $\mathcal{M}$ has now to be found. Substituting in (3) for $\mathcal{M}_x$ and $\mathcal{M}_y$ from (18), we have

$$\dot{\mathcal{M}}_z + \mathcal{M}_z/T_1 = \mathcal{M}_0/T_1. \qquad (26)$$

Now from (19) we find that

$$\dot{\mathcal{M}} = \mathcal{M}_z \dot{\mathcal{M}}_z / \mathcal{M} + \mathcal{M}_1 \dot{\mathcal{M}}_1 / \mathcal{M}. \tag{27}$$

Using (20) and (26) for $\dot{\mathcal{M}}_1$ and $\dot{\mathcal{M}}_z$ respectively, and using (24), we get

$$\dot{\mathcal{M}} + \frac{\mathcal{M}}{T_1} \frac{(\delta^2 + T_1/T_2)}{(1 + \delta^2)} = \frac{\mathcal{M}_0 \delta}{T_1 (1 + \delta^2)^{\frac{1}{2}}}. \tag{28}$$

Equation (28) is a straightforward linear first-order differential equation in $\mathcal{M}$, whose solution is

$$\mathcal{M}(t) = \int_{-\infty}^{t} dt' \frac{\mathcal{M}_0(t')\delta(t')}{T_1(1 + \delta^2(t'))^{\frac{1}{2}}} \exp\left[\int_{t}^{t'} \frac{\delta^2(t'') + T_1/T_2}{T_1(1 + \delta^2(t''))} dt'' \right]. \tag{29}$$

where $\mathcal{M}_0(t') = \chi_0 H_0(t')$. This solution is used in §5.5.1.

TABLE OF NUCLEAR MOMENTS

This table is based on one compiled by Dr A. L. Bloom, and reproduced by kind permission of Varian Associates, Palo Alto, California. The spin I of each nucleus is given in units of $\hbar$; the magnetic moment μ, without diamagnetic correction, in units of the nuclear magneton; the electric quadrupole moment Q in units of 10^{-24} cm^2; and the Larmor frequency in a field of 10,000 gauss in Mc/s. References are given to determinations of I and μ which have used the nuclear magnetic resonance method. An asterisk * indicates that the nucleus is radioactive.

Nucleus	I	μ	Larmor frequency in 10 kG.	Q	References (the numbers refer to the bibliography below this table)
^{1}n*	$\frac{1}{2}$	$-1\cdot91315$	$29\cdot1681$	—	1, 2, 3
^{1}H	$\frac{1}{2}$	$2\cdot79268$	$42\cdot5776$	—	3, 4, 5, 6, 79
^{2}H	1	$0\cdot85738$	$6\cdot5359$	$2\cdot77 \times 10^{-3}$	1, 7, 8, 9, 10, 11, 12, 13, 14, 73, 76
^{3}H*	$\frac{1}{2}$	$2\cdot9787$	$45\cdot414$	—	15, 16
^{3}He	$\frac{1}{2}$	$-2\cdot1274$	$32\cdot435$	—	17
^{6}Li	1	$0\cdot82191$	$6\cdot265$	$4\cdot6 \times 10^{-4}$	18, 75
^{7}Li	$\frac{3}{2}$	$3\cdot2560$	$16\cdot547$	$-4\cdot2 \times 10^{-2}$	10, 11, 19, 20, 21, 74, 75, 76
^{9}Be	$\frac{3}{2}$	$-1\cdot1774$	$5\cdot983$	2×10^{-2}	22, 23, 24, 25, 26, 27
^{10}B	3	$1\cdot8006$	$4\cdot575$	$0\cdot111$	20, 28
^{11}B	$\frac{3}{2}$	$2\cdot6880$	$13\cdot660$	$3\cdot55 \times 10^{-2}$	11, 20, 25, 29, 30, 76
^{13}C	$\frac{1}{2}$	$0\cdot70216$	$10\cdot705$	—	31
^{14}N	1	$0\cdot40357$	$3\cdot076$	$7\cdot1 \times 10^{-2}$	28, 32, 33
^{15}N	$\frac{1}{2}$	$-2\cdot8304$	$4\cdot315$	—	32, 33
^{17}O	$\frac{5}{2}$	$-1\cdot8930$	$5\cdot772$	-4×10^{-3}	34
^{19}F	$\frac{1}{2}$	$2\cdot6272$	$40\cdot055$	—	10, 11, 25, 31, 35, 36, 37, 76, 78
^{21}Ne	$\frac{3}{2}$	—	—	—	—
^{22}Na*	3	$1\cdot745$	$\cdot434$	—	—
^{23}Na	$\frac{3}{2}$	$2\cdot2161$	$1\cdot262$	$0\cdot1$	11, 20, 21, 25, 38, 74, 76
^{24}Na*	4	$1\cdot69$	$3\cdot22$	—	—
^{25}Mg	$\frac{5}{2}$	$-0\cdot85470$	$2\cdot606$	—	27
^{27}Al	$\frac{5}{2}$	$3\cdot6385$	$11\cdot094$	$0\cdot149$	11, 20, 21, 25, 35, 39, 74, 76, 80
^{29}Si	$\frac{1}{2}$	$-0\cdot55477$	$8\cdot458$	—	40, 41, 42, 83, 86
^{31}P	$\frac{1}{2}$	$1\cdot1305$	$17\cdot236$	—	20, 21, 24, 25, 43

Nucleus	I	μ	Larmor frequency in 10 kG.	Q	References (the numbers refer to the bibliography below this table)
^{33}S	3/2	0·64274	3·266	$-6\cdot4 \times 10^{-2}$	42, 44
^{35}S*	3/2	1·00	5·08	$4\cdot5 \times 10^{-2}$	—
^{35}Cl	3/2	0·82088	4·172	$-7\cdot97 \times 10^{-2}$	20, 24, 28, 33, 45
^{36}Cl*	2	1·2839	4·893	$-1\cdot68 \times 10^{-2}$	89
^{37}Cl	3/2	0·68328	3·472	$-6\cdot21 \times 10^{-2}$	18, 28, 32, 33
^{39}K	3/2	0·39094	1·987	0·14	46
^{40}K*	4	−1·296	2·470	—	—
^{41}K	3/2	0·21453	1·090	—	81
^{42}K*	2	−1·14	4·34	—	—
^{43}Ca	7/2	−1·3153	2·865	—	47
^{45}Sc	7/2	4·7491	10·344	—	33, 48, 49, 50
^{47}Ti	5/2	−0·78711	2·400	—	51, 85
^{49}Ti	7/2	−1·1023	2·401	—	51, 85
^{49}V*	7/2	4·68	10·2	—	—
^{50}V	6	3·3413	4·245	—	45
^{51}V	7/2	5·1392	11·193	0·3	25, 33, 52, 88
^{53}Cr	3/2	−0·47354	2·406	—	53, 84
^{53}Mn*	7/2	5·050	11·00	—	—
^{55}Mn	5/2	3·4610	10·553	0·5	25, 32, 33
^{56}Co*	4	3·855	7·347	—	—
^{57}Co*	7/2	4·65	10·1	—	—
^{58}Co*	2	4·052	15·44	—	—
^{59}Co	7/2	4·6388	10·103	0·5	32, 33
^{60}Co*	5	3·800	5·793	—	—
^{63}Cu	3/2	2·2206	11·285	−0·15	11, 20, 25, 54
^{64}Cu*	1	0·40	3·0	—	—
^{65}Cu	3/2	2·3789	12·090	−0·14	11, 20, 25, 54
^{67}Zn	5/2	0·8735	2·664	0·18	42, 55
^{69}Ga	3/2	2·0107	10·219	0·2318	43
^{71}Ga	3/2	2·5549	12·984	0·1461	43
^{73}Ge	9/2	−0·8768	1·485	−0·2	85
^{75}As	3/2	1·4348	7·292	0·3	28, 42, 51, 56
^{77}Se	1/2	0·5333	8·131	—	42, 57, 77
^{79}Se*	7/2	−1·015	2·211	0·9	—
^{79}Br	3/2	2·0990	10·667	0·33	11, 25, 58
^{81}Br	3/2	2·2626	11·499	0·28	11, 20, 25, 58
^{83}Kr	9/2	−0·968	1·64	0·15	—
^{85}Kr*	9/2	1·00	1·70	0·25	—
^{81}Rb*	3/2	2·00	10·2	—	—
^{85}Rb	5/2	1·3482	4·111	0·31	20, 24, 45, 59, 60
^{86}Rb*	2	(−)1·7	6·5	—	—
^{87}Rb	3/2	2·7414	13·932	0·15	11, 20, 25, 59, 60
^{87}Sr	9/2	−1·0893	1·845	—	84
^{89}Y	1/2	−0·1368	2·086	—	81
^{91}Zr	5/2	−1·298	3·958	—	90
^{93}Nb	9/2	6·1435	10·407	−0·2	25, 61
^{95}Mo	5/2	−0·9099	2·774	—	33
^{97}Mo	5/2	−0·9290	2·833	—	33
^{99}Tc*	9/2	5·6572	9·583	0·3	62
^{99}Ru	5/2	−0·63	1·9	—	—

Nucleus	I	μ	Larmor frequency in 10 kG.	Q	References (the numbers refer to the bibliography below this table)
^{101}Ru	$\frac{5}{2}$	$-0\cdot69$	$2\cdot1$	—	—
^{103}Rh	$\frac{1}{2}$	$-0\cdot0879$	$1\cdot340$	—	89
^{105}Pd	$\frac{5}{2}$	$-0\cdot57$	$1\cdot74$	—	—
^{107}Ag	$\frac{1}{2}$	$-0\cdot1130$	$1\cdot723$	—	81, 82
^{109}Ag	$\frac{1}{2}$	$-0\cdot1299$	$1\cdot981$	—	81, 82
^{111}Ag*	$\frac{1}{2}$	$-0\cdot145$	$2\cdot21$	—	—
^{111}Cd	$\frac{1}{2}$	$-0\cdot5922$	$9\cdot028$	—	63, 64
^{113}Cd	$\frac{1}{2}$	$-0\cdot6195$	$9\cdot444$	—	63, 64
^{113}In	$\frac{9}{2}$	$5\cdot4956$	$9\cdot310$	$1\cdot144$	28, 33
^{115}In*	$\frac{9}{2}$	$5\cdot5072$	$9\cdot329$	$1\cdot161$	28, 33
^{115}Sn	$\frac{1}{2}$	$-0\cdot9132$	$13\cdot92$	—	63, 64
^{117}Sn	$\frac{1}{2}$	$-0\cdot9949$	$15\cdot17$	—	64, 65
^{119}Sn	$\frac{1}{2}$	$-1\cdot0408$	$15\cdot87$	—	64, 65
^{121}Sb	$\frac{5}{2}$	$3\cdot3417$	$10\cdot19$	$-0\cdot53$	33, 48, 66, 67
^{123}Sb	$\frac{7}{2}$	$2\cdot5334$	$5\cdot518$	$-0\cdot68$	33, 48, 66
^{123}Te	$\frac{1}{2}$	$-0\cdot7319$	$11\cdot16$	—	41, 42
^{125}Te	$\frac{1}{2}$	$-0\cdot8824$	$13\cdot45$	—	41, 42
^{127}I	$\frac{5}{2}$	$2\cdot7940$	$8\cdot519$	$-0\cdot69$	11, 25, 43, 59, 68
^{129}I*	$\frac{7}{2}$	$2\cdot6030$	$5\cdot669$	$-0\cdot43$	68
^{129}Xe	$\frac{1}{2}$	$-0\cdot7725$	$11\cdot78$	—	33, 48, 87
^{131}Xe	$\frac{3}{2}$	$0\cdot6868$	$3\cdot490$	$-0\cdot12$	87
^{127}Cs*	$\frac{1}{2}$	$1\cdot41$	$21\cdot5$	—	—
^{129}Cs*	$\frac{1}{2}$	$1\cdot47$	$22\cdot4$	—	—
^{131}Cs*	$\frac{5}{2}$	$3\cdot48$	$10\cdot6$	—	—
^{133}Cs	$\frac{7}{2}$	$2\cdot5642$	$5\cdot585$	-3×10^{-3}	20, 22, 24, 25
^{134}Cs*	4	$2\cdot96$	$5\cdot64$	—	—
^{135}Cs*	$\frac{7}{2}$	$2\cdot727$	$5\cdot94$	—	—
^{137}Cs*	$\frac{7}{2}$	$2\cdot84$	$6\cdot18$	—	—
^{135}Ba	$\frac{3}{2}$	$0\cdot8323$	$4\cdot230$	—	91
^{137}Ba	$\frac{3}{2}$	$0\cdot9311$	$4\cdot732$	—	91
^{138}La*	5	$3\cdot684$	$5\cdot617$	$2\cdot7$	92
^{139}La	$\frac{7}{2}$	$2\cdot7614$	$6\cdot014$	$0\cdot6$	24, 25, 69
^{141}Ce*	$\frac{7}{2}$	$0\cdot16$	$0\cdot35$	—	—
^{141}Pr	$\frac{5}{2}$	$3\cdot92$	$11\cdot95$	$-5\cdot4\times10^{-2}$	—
^{143}Nd	$\frac{7}{2}$	$-1\cdot03$	$2\cdot24$	$<1\cdot2$	—
^{145}Nd	$\frac{7}{2}$	$-0\cdot64$	$1\cdot4$	$<1\cdot2$	—
^{147}Nd*	$\frac{5}{2}$	$0\cdot22$	$0\cdot37$	—	—
^{147}Sm	$\frac{7}{2}$	$-0\cdot83$	$1\cdot8$	$0\cdot72$	—
^{149}Sm	$\frac{7}{2}$	$-0\cdot68$	$1\cdot5$	$0\cdot72$	—
^{151}Eu	$\frac{5}{2}$	$3\cdot4$	10	$\sim1\cdot2$	—
^{153}Eu	$\frac{5}{2}$	$1\cdot5$	$4\cdot6$	$\sim2\cdot5$	—
^{154}Eu*	3	$2\cdot0$	$5\cdot1$	—	—
^{155}Gd	$\frac{3}{2}$	$-0\cdot24$	$1\cdot2$	$1\cdot1$	—
^{157}Gd	$\frac{3}{2}$	$-0\cdot32$	$1\cdot6$	$1\cdot0$	—
^{159}Tb	$\frac{3}{2}$	$1\cdot52$	$7\cdot72$	—	—
^{161}Dy	$\frac{5}{2}$	$0\cdot38$	$1\cdot2$	—	—
^{163}Dy	$\frac{5}{2}$	$0\cdot53$	$1\cdot6$	—	—
^{165}Ho	$\frac{7}{2}$	$3\cdot29$	$7\cdot17$	2	—
^{167}Er	$\frac{7}{2}$	$0\cdot48$	$1\cdot04$	~10	—
^{169}Tm	$\frac{1}{2}$	$-0\cdot20$	$3\cdot05$	—	—

Nucleus	I	μ	Larmor frequency in 10 kG.	Q	References (the numbers refer to the bibliography below this table)
^{171}Yb	$\frac{1}{2}$	0·43	6·6	—	—
^{173}Yb	$\frac{5}{2}$	−0·60	1·8	3·9	—
^{175}Lu	$\frac{7}{2}$	2·6	5·7	5·8	—
^{176}Lu*	$\geqslant 7$	4·2	—	6–8	—
^{177}Hf	$\frac{7}{2}$	0·61	1·3	3	—
^{179}Hf	$\frac{9}{2}$	−0·47	0·80	3	—
^{181}Ta	$\frac{7}{2}$	2·1	4·6	6·0	—
^{183}W	$\frac{1}{2}$	0·115	1·75	—	89
^{185}Re	$\frac{5}{2}$	3·1437	9·586	2·8	27
^{187}Re	$\frac{5}{2}$	3·1760	9·684	2·6	27
^{187}Os	$\frac{1}{2}$	∼0·12	∼1·8	—	—
^{189}Os	$\frac{3}{2}$	0·6506	3·307	2·0	93
^{191}Ir	$\frac{3}{2}$	0·16	0·813	∼1·2	—
^{193}Ir	$\frac{3}{2}$	0·17	0·86	∼1·0	—
^{195}Pt	$\frac{1}{2}$	0·6004	9·153	—	33, 63
^{197}Au	$\frac{3}{2}$	0·136	0·691	0·56	—
^{198}Au*	2	0·50	1·9	—	—
^{199}Au*	$\frac{3}{2}$	0·24	1·2	—	—
^{197}Hg*	$\frac{1}{2}$	0·52	7·9	—	—
^{199}Hg	$\frac{1}{2}$	0·499	7·61	—	33, 63
^{201}Hg	$\frac{3}{2}$	−0·607	3·08	0·5	—
^{203}Tl	$\frac{1}{2}$	1·5960	24·33	—	25, 31, 64, 71, 72
^{205}Tl	$\frac{1}{2}$	1·6114	24·57	—	25, 31, 64, 70, 71, 72
^{207}Pb	$\frac{1}{2}$	0·5837	8·899	—	64, 65
^{209}Bi	$\frac{9}{2}$	4·0389	6·842	−0·4	28, 33, 48
^{227}Ac*	$\frac{3}{2}$	1·1	5·6	−1·7	—
^{233}U*	$\frac{5}{2}$	0·51	1·5	0·34	—
^{235}U*	$\frac{7}{2}$	0·51	1·1	4·0	—
^{237}Np*	$\frac{5}{2}$	6±2·5	20	—	—
^{239}Pu*	$\frac{1}{2}$	0·4	6·1	—	—
^{241}Pu*	$\frac{5}{2}$	1·4	4·3	—	—
^{241}Am*	$\frac{5}{2}$	1·4	4·3	4·9	—
^{243}Am*	$\frac{5}{2}$	1·4	4·3	4·9	—

1. Arnold and Roberts (1946, 1947). 2. Bloch, Nicodemus and Staub (1948). 3. Rogers and Staub (1949); Staub and Rogers (1950). 4. Thomas, Driscoll and Hipple (1949, 1950 a, b). 5. Bloch and Jeffries (1950); Jeffries (1951); Hipple, Sommer and Thomas (1949); Sommer, Thomas and Hipple (1950, 1951). 6. Gardner and Purcell (1949); Gardner (1951); Nelson (unpublished). 7. Roberts (1947 b). 8. Bitter, Alpert, Nagle and Poss (1947). 9. Bloch, Levinthal and Packard (1947). 10. Siegbahn and Lindström (1949 a, b). 11. Zimmerman and Williams (1949 c). 12. Levinthal (1950). 13. Lindström (1950, 1951 a, b, c). 14. Smaller, Yasaitis and Anderson (1951); Smaller (1951).

15. Anderson and Novick (1947). 16. Bloch, Graves, Packard and Spence (1947 a, b). 17. Anderson and Novick (1948); Anderson, H. L. (1949). 18. Watkins and Pound (1951). 19. Bolle, Puppi and Zanotelli (1946). 20. Bitter (1949 a). 21. Kanda, Masuda, Kusaka, Yamagata and Itoh (1952). 22. Zimmerman and Williams (1949 b). 23. Dickinson and Wimett (1949). 24. Chambers and Williams (1949). 25. Sheriff and Williams (1951). 26. Gutowsky, McClure and Hoffman (1951); Schuster and Pake (1951 b) Hatton, Rollin and Seymour (1951). 27. Alder and Yu (1951 b). 28. Ting and Williams (1953). 29. Zimmerman and Williams (1948). 30. Anderson, D. A. (1949). 31. Poss (1949). 32, 33. Proctor and Yu (1950 a, 1951). 34. Alder and Yu (1951 a). 35. Guptill, Archibald and Warren (1950). 36. Kanda, Masuda, Kusaka, Yamagata and Itoh (1951); Kanda (1952). 37. Béné (1951). 38. Bolle and Zanotelli (1948 a). 39. Zimmerman and Williams (1949 a). 40. Hatton, Rollin and Seymour (1951). 41. Dharmatti and Weaver (1951 c). 42. Weaver (1953). 43. Pound (1948 b). 44. Dharmatti and Weaver (1951 a). 45. Walchli, Leyshon and Scheitlin (1952). 46. Collins (1950 b). 47. Jeffries (1953 a). 48. Proctor and Yu (1950 b). 49. Sheriff and Williams (1950); Ramsey (1950 c). 50. Hunten (1950, 1951). 51. Jeffries, Löliger and Staub (1951, 1952). 52. Knight and Cohen (1949). 53. Alder and Halbach (1953). 54. Pound (1948 a). 55. Dharmatti and Weaver (1952 a). 56. Dharmatti and Weaver (1951 b). 57. Dharmatti and Weaver (1952 b). 58. Pound (1947 b). 59. Yasaitis and Smaller (1951). 60. Adams, Wimett and Bitter (1951). 61. Sheriff, Chambers and Williams (1950). 62. Walchli, Livingston and Martin (1952). 63. Proctor and Yu (1949). 64. Proctor (1950). 65. Proctor (1949 b). 66. Cohen, Knight, Wentink and Koski (1950). 67. Collins (1950 a). 68. Walchli, Livingston and Hebert (1951). 69. Dickinson (1949). 70. Poss (1947). 71. Proctor (1949 a). 72. Gutowsky and McGarvey (1953 a). 73. Wimett (1953 b). 74. Pound (1950). 75. Schuster and Pake (1951 a). 76. Lindström (1951 b). 77. Walchli (1953 a). 78. Bloembergen (1948). 79. Chiarotti and Giulotto (1951). 80. Petch, Cranna and Volkoff (1953). 81. Brun, Oeser, Staub and Telschow (1954 a). 82. Sogo and Jeffries (1954). 83. Ogg and Ray (1954). 84. Jeffries and Sogo (1953). 85. Jeffries (1953 b).

86. Williams, McCall and Gutowsky (1954). 87. Brun, Oeser, Staub and Telschow (1954*b*). 88. Walchli and Morgan (1952). 89. Sogo and Jeffries, *Phys. Rev.* **98**, 1316 (1955). 90. Brun, Oeser and Staub, *Phys. Rev.* **105**, 1929 (1957). 91. Walchli and Rowland, *Phys. Rev.* **102**, 1334 (1956). 92. Sogo and Jeffries, *Phys. Rev.* **99**, 613 (1955). 93. Loeliger and Sarles, *Phys. Rev.* **95**, 291 (1954).

NUCLEAR MAGNETIC RELAXATION IN LIQUIDS†

The classical expression for the potential energy V_{ij} of magnetic interaction between two magnetic dipoles of moment μ_i and μ_j joined by a vector $\mathbf{r}_{ij}$ is

$$V_{ij} = \left(\frac{\mu_i \cdot \mu_j}{r_{ij}^3} \right) - \frac{3(\mu_i \cdot \mathbf{r}_{ij})(\mu_j \cdot \mathbf{r}_{ij})}{r_{ij}^5}. \tag{1}$$

If μ_i is the magnetic moment of the ith nucleus, and μ_j is the moment of its jth neighbour, then expression (1), treated as an operator, and summed over all neighbours, is the dipole-dipole interaction term in the Hamiltonian. Writing $\mu = \gamma \hbar \mathbf{I}$, expression (1) for a system of identical nuclear magnetic dipoles becomes

$$V_{ij} = \gamma^2 \hbar^2 r_{ij}^{-3} [(\mathbf{I}_i \cdot \mathbf{I}_j) - 3(\mathbf{I}_i \cdot \mathbf{p}_{ij})(\mathbf{I}_j \cdot \mathbf{p}_{ij})], \tag{2}$$

where $\mathbf{p}_{ij}$ is a unit vector parallel to $\mathbf{r}_{ij}$. If $\lambda_1, \lambda_2, \lambda_3$ are unit vectors parallel to Cartesian axes x, y and z, where the z axis is taken parallel to $\mathbf{H}_0$, we may write

$$\mathbf{I}_i = \lambda_1 I_{x_i} + \lambda_2 I_{y_i} + \lambda_3 I_{z_i}, \tag{3}$$

with a similar equation for $\mathbf{I}_j$. Similarly, we have

$$\mathbf{p}_{ij} = \lambda_1 \epsilon_1 + \lambda_2 \epsilon_2 + \lambda_3 \epsilon_3, \tag{4}$$

where ϵ_1, ϵ_2, ϵ_3 are the direction cosines of $\mathbf{p}_{ij}$ (and therefore also of $\mathbf{r}_{ij}$). Forming the scalar products in (2) with the aid of (3) and (4) we get

$$V_{ij} = \gamma^2 \hbar^2 r_{ij}^{-3} [I_{x_i} I_{x_j} (1 - 3\epsilon_1^2) + I_{y_i} I_{y_j} (1 - 3\epsilon_2^2) + I_{z_i} I_{z_j} (1 - 3\epsilon_3^2)$$
$$- 3(I_{x_i} I_{y_j} + I_{x_j} I_{y_i}) \epsilon_1 \epsilon_2 - 3(I_{y_i} I_{z_j} + I_{y_j} I_{z_i}) \epsilon_2 \epsilon_3$$
$$- 3(I_{z_i} I_{x_j} + I_{z_j} I_{x_i}) \epsilon_3 \epsilon_1]. \tag{5}$$

We now transform to spherical polar co-ordinates, with polar angle θ_{ij} between $\mathbf{r}_{ij}$ and $\mathbf{H}_0$, and azimuthal angle ϕ_{ij}. Consequently

$$\epsilon_1 = \sin \theta_{ij} \cos \phi_{ij}, \quad \epsilon_2 = \sin \theta_{ij} \sin \phi_{ij}, \quad \epsilon_3 = \cos \theta_{ij}. \tag{6}$$

If we now write $\cos \phi_{ij}$ as $\frac{1}{2}\{\exp(i\phi_{ij}) + \exp(-i\phi_{ij})\}$ and $\sin \phi_{ij}$ as

† Bloembergen, Purcell and Pound (1948).

$\frac{1}{2i}\{\exp(i\phi_{ij}) - \exp(-i\phi_{ij})\}$ and substitute (6) in (5) and collect terms, we find

$$V_{ij} = \gamma^2\hbar^2 r_{ij}^{-3}(A + B + C + D + E + F), \qquad (7)$$

where

$$
\left.
\begin{aligned}
A &= I_{z_i}I_{z_j}(1 - 3\cos^2\theta_{ij}), \\
B &= -\tfrac{1}{4}[(I_{x_i} - iI_{y_i})(I_{x_j} + iI_{y_j}) \\
&\quad + (I_{x_i} + iI_{y_i})(I_{x_j} - iI_{y_j})](1 - 3\cos^2\theta_{ij}), \\
C &= -\tfrac{3}{2}[(I_{x_i} + iI_{y_i})I_{z_j} + (I_{x_j} + iI_{y_j})I_{z_i}]\sin\theta_{ij}\cos\theta_{ij}\exp(-i\phi_{ij}), \\
D &= -\tfrac{3}{2}[(I_{x_i} - iI_{y_i})I_{z_j} + (I_{x_j} - iI_{y_j})I_{z_i}]\sin\theta_{ij}\cos\theta_{ij}\exp(i\phi_{ij}), \\
E &= -\tfrac{3}{4}(I_{x_i} + iI_{y_i})(I_{x_j} + iI_{y_j})\sin^2\theta_{ij}\exp(-2i\phi_{ij}), \\
F &= -\tfrac{3}{4}(I_{x_i} - iI_{y_i})(I_{x_j} - iI_{y_j})\sin^2\theta_{ij}\exp(2i\phi_{ij}).
\end{aligned}
\right\} \quad (8)
$$

In a rigid system terms A and B give rise to secular perturbations. On the other hand, terms C to F produce only periodic perturbations of small amplitude in the precessional motion of nucleus i. This may be seen if one transforms to co-ordinates rotating about $\mathbf{H}_0$ with the precessional angular frequency γH_0, replacing ϕ_{ij} with $(\phi'_{ij} \pm \gamma H_0 t)$. For a liquid, however, the factors involving r_{ij}, θ_{ij} and ϕ_{ij} are time-dependent. Thus if the intensity of the Fourier spectrum of the term in D is finite at frequency $\nu_0 = \gamma H_0/2\pi$, the time factor cancels in the rotating system and a secular perturbation is obtained, producing transitions involving a change $\Delta m_i = -1$ in the quantum number of nucleus i. Similarly, the Fourier component of opposite sense (frequency $-\nu_0$) allows C to give transitions involving a change $\Delta m_i = +1$. Furthermore Fourier components at frequencies $\pm 2\nu_0$ enable E and F to produce transitions. The physical significance of these transitions is discussed in §5.2.

We now have to find the intensity of the Fourier spectra of the position functions which enter the terms A to F, and which are defined as

$$
\left.
\begin{aligned}
Y_{0j} &= (1 - 3\cos^2\theta_{ij})r_{ij}^{-3}, \\
Y_{1j} &= \sin\theta_{ij}\cos\theta_{ij}\exp(i\phi_{ij})r_{ij}^{-3}, \\
Y_{2j} &= \sin^2\theta_{ij}\exp(2i\phi_{ij})r_{ij}^{-3}.
\end{aligned}
\right\} \quad (9)
$$

To each of these randomly varying functions $Y(t)$ we can ascribe a correlation function $K(\tau)$ defined as

$$K(\tau) = \overline{Y(t)\,Y^*(t+\tau)}. \qquad (10)$$

Q

For $\tau = 0$, we have

$$K(0) = \overline{|\,Y(t)\,|}^2. \tag{11}$$

The spectral density $J(\nu)$ of the random function Y is given by the Fourier transform of $K(\tau)$ (Wang and Uhlenbeck, 1945),

$$J(\nu) = \int_{-\infty}^{\infty} K(\tau) \exp(2\pi i\nu\tau)d\tau. \tag{12}$$

For simplicity the correlation function $K(\tau)$ is assumed to be given by

$$K(\tau) = K(0)\exp(-\,|\,\tau\,|/\tau_c), \tag{13}$$

where τ_c is called the correlation time. There is in general a different correlation time τ_{cj} for each nuclear neighbour j of the nucleus i. Substituting (13) in (12), carrying out the integration, and using (11), we find for one neighbour j that the spectral density contributed at i is

$$J_j(\nu) = \overline{|\,Y_j\,|}^2 \, 2\tau_{cj}(1 + 4\pi^2\nu^2\tau_{cj}^2)^{-1}. \tag{14}$$

The total spectral density $J(\nu)$ for all neighbours j is therefore

$$J(\nu) = \sum_j J_j(\nu) = \sum_j \overline{|\,Y_j\,|}^2 \, 2\tau_{cj}(1 + 4\pi^2\nu^2\tau_{cj}^2)^{-1}. \tag{15}$$

We note therefore by integration that

$$\int_{-\infty}^{\infty} J(\nu)d\nu = \sum_j \overline{|\,Y_j\,|}^2. \tag{16}$$

In all these equations (10)–(16) subscripts 0, 1, 2 must be inserted when we are considering one particular member of the group Y_{0j}, Y_{1j}, Y_{2j}.

Having now found the spectral intensities of the position functions entering in the terms C to F of (7), we now have to find the probability of the transitions which they bring about. Let us first consider the contribution of term D of (7) to the probability $W_{m_i \to m_i-1}$ of transitions in which the magnetic quantum number m_i of the ith nucleus changes to $m_i - 1$; the part of the magnetic interaction operator (7) concerned is

$$(V_{ij})_D = -\tfrac{3}{2}\gamma^2\hbar^2[(I_{x_i} - iI_{y_i})I_{z_j} + (I_{x_j} - iI_{y_j})I_{z_i}]Y_{1j}. \tag{17}$$

From first-order perturbation theory the contribution this term makes to the transition probability is

$$\frac{1}{\hbar^2}(\tfrac{3}{2}\gamma^2\hbar^2)^2\,|\,\langle m_i\,|\,(I_{x_i} - iI_{y_i})I_{z_j} + (I_{x_j} - iI_{y_j})I_{z_i}\,|\,m_i - 1\rangle\,|^2\,J_{1j}(\nu_0).\,(18)$$

Remembering that I_{z_i} and I_{z_j} are just m_i and m_j respectively, and recalling (see Condon and Shortley, 1935, §3[3]) that

$$\langle m_i \mid I_{x_i} - iI_{v_i} \mid m_i - 1 \rangle = \sqrt{\{(I+m_i)(I-m_i+1)\}}, \qquad (19)$$

we find that (18) becomes

$$\tfrac{9}{4}\gamma^4\hbar^2 m_j^2 (I+m_i)(I-m_i+1) J_{1j}(\nu_0). \qquad (20)$$

This expression must now be summed over all the neighbours j. Since the magnetic energy levels designated by different m_j are closely spaced compared with kT, we may assume that the neighbouring nuclei are evenly distributed among the $(2I+1)$ levels. The factor m_j^2 may therefore be replaced by its mean value $\tfrac{1}{3}I(I+1)$ when the sum is formed. The sum of (20) over all neighbours j thus gives

$$\tfrac{3}{4}\gamma^4\hbar^2 I(I+1)(I+m_i)(I-m_i+1) J_1(\nu_0). \qquad (21)$$

A contribution to $W_{m_i \to m_i-1}$ is also forthcoming from term F of (7). Carrying out a similar calculation for term F to that for term D, we find an expression like (21), but with $J_1(\nu_0)$ replaced by $\tfrac{1}{2}J_2(2\nu_0)$. The total transition probability is therefore

$$W_{m_i \to m_i-1} = \tfrac{3}{4}\gamma^4\hbar^2 I(I+1)(I+m_i)(I-m_i+1)[J_1(\nu_0)+\tfrac{1}{2}J_2(2\nu_0)]. \quad (22)$$

Similarly from terms C and E we find

$$W_{m_i \to m_i+1} = \tfrac{3}{4}\gamma^4\hbar^2 I(I+1)(I-m_i)(I+m_i+1)[J_1(-\nu_0)+\tfrac{1}{2}J_2(-2\nu_0)]. \quad (23)$$

If we apply (22) to the case of nuclei with $I=\tfrac{1}{2}$, then we can only have $m_i = +\tfrac{1}{2}$, while for (23) we can only have $m_i = -\tfrac{1}{2}$, and both equations become

$$W = \tfrac{3}{4}\gamma^4\hbar^2 I(I+1)[J_1(\nu_0)+\tfrac{1}{2}J_2(2\nu_0)], \qquad (24)$$

remembering that the $J(\nu)$ are even functions. The spin-lattice relaxation time is then given by $T_1 = 1/2W$ (equation (2.11)). We mentioned in §2.2 that for $I > \tfrac{1}{2}$, the approach of spin system and lattice to thermal equilibrium is exponential provided that the populations of the successive levels are related by the same factor, so that a spin temperature may be defined. Bloembergen (1948) has shown, using (22) and (23), that if this condition is met, then

T_1 is still given by $1/2W$, where W is given by (24). Hence we have

$$\frac{1}{T_1} = 2W = \tfrac{3}{2}\gamma^4\hbar^2 I(I+1)\left[J_1(\nu_0) + \tfrac{1}{2}J_2(2\nu_0)\right] \tag{25}$$

$$= \tfrac{3}{2}\gamma^4\hbar^2 I(I+1)\sum_j\left[\frac{2\tau_{cj}\;\overline{\left|Y_{1j}\right|^2}}{1 + 4\pi^2\nu_0^2\tau_{cj}^2} + \frac{\tau_{cj}\;\overline{\left|Y_{2j}\right|^2}}{1 + 16\pi^2\nu_0^2\tau_{cj}^2}\right], \tag{26}$$

using (15).

We have so far assumed that the magnetic dipoles are identical. If, however, some of the neighbouring nuclei, of which k is typical, are different, it may readily be seen that the following more general term must be added to expression (25) for $1/T_1$:

$$\tfrac{3}{4}\gamma_i^2\hbar^2 \sum_k \gamma_k^2 I_k(I_k+1)\left[J_{1k}(\nu_{0i}) + \tfrac{1}{2}J_{2k}(\nu_{0i}+\nu_{0k})\right]. \tag{27}$$

Expression (27) may also be used for paramagnetic ion neighbours if we substitute the square of the magnetic moment of the ion for $\gamma_k^2\hbar^2 I_k(I_k+1)$. The validity of the expression assumes, however, that the spatial quantization of the ion does not change in a time of the order of the correlation times τ_c, and since the relaxation times for paramagnetic ions may be as small as 10^{-9} sec., this condition may not always be met.

Note added in proof

As is mentioned on p. 150 Kubo and Tomita (1954) have re-examined and revised some of the equations developed by Bloembergen, Purcell and Pound (1948) which are quoted in this appendix.

NON-METALLIC SOLIDS INVESTIGATED BY THE NUCLEAR MAGNETIC RESONANCE METHOD

This table lists solids in which nuclear magnetic resonances have been reported. Inorganic compounds are arranged in order of increasing atomic number of the first element in the conventional chemical formula. An asterisk * against a reference in the final column indicates that information concerning the crystal structure has been obtained; a dagger † indicates that evidence has been found of molecular motion in the crystal.

Inorganic

Solid	Resonant nuclei	References
H_2	^1H	1, 2†, 3†, 4†, 5†, 86
HD	^1H	5
D_2	^2H	4†, 5†
HCl	^1H	6, 7†
HBr	^1H	7†
HI	^1H	7†
H_2O	^1H	8, 9†, 10, 11
H_2S	^1H	7†
H_2Se	^1H	7†
$HNO_3.H_2O$	^1H	12*
$H_2SO_4.H_2O$	^1H	12*
$H_2SO_4.2H_2O$	^1H	13*
$HClO_4.H_2O$	^1H	12*†, 14*†, 15*
$HClO_4.2H_2O$	^1H	13*
$H_2SeO_4.H_2O$	^1H	13*
$H_2PtCl_6.2H_2O$	^1H	13*
LiF	^7Li, ^{19}F	1, 2, 8, 16, 17, 18, 19
LiCl	^7Li	2, 16
Li_2CO_3	^7Li	20
$LiNO_3$	^7Li	20
$Li_2SO_4.H_2O$	^1H, ^7Li	20, 21*
LiOH	^1H	8
$LiAl(SiO_3)_2$ (spodumene)	^6Li, ^7Li, ^{27}Al	22, 23
$LiAlH_4$	^1H	24
$LiBH_4$	^1H	24
BeF_2	^{19}F	25
$BeAl_2O_4$ (chrysoberyl)	^9Be	26

Solid	Resonant nuclei	References
$Be_3Al_2Si_6O_{18}$	^9Be, ^{27}Al, ^{29}Si	27
B_2H_6	^1H	28*
NH_3	^1H	29†
NH_4F	^1H	28*, 84†
NH_4Cl	^1H	7, 8, 11†, 28*, 30, 31†, 32*†, 33†, 83*†, 84*†
NH_4Br	^1H	11†, 28*, 32*†, 33†
NH_4I	^1H	32†, 33†
NH_4NO_3	^1H	33†
$(NH_4)_2SO_4$	^1H	30, 33†
NH_4CNS	^1H	33†
NH_4IO_3	^1H	33†
$NH_4H_2PO_4$	^1H	34†, 84
$NH_4H_2AsO_4$	^1H	34†
$(NH_4)_2BeF_4$	^{19}F	25
$(NH_4)_2CuCl_4.2H_2O$	^1H	35*
N_2H_5Cl	^1H	36
$N_2H_6Cl_2$	^1H	36†
$N_2H_6F_2$	^1H, ^{19}F	36, 81*
$N_2H_6SO_4$	^1H	36*†
$N_2H_6(NO_3)_2$	^1H	36*†
$N_2H_6C_2O_4$	^1H	36*†
$(N_2H_5)_2SO_4$	^1H	36*†
$(N_2H_5)_2C_2O_4$	^1H	36*†
$N_2H_5BF_4$	^1H	36*
NaF	^{19}F, ^{23}Na	1, 16, 25
$NaBr$	^{23}Na, ^{79}Br, ^{81}Br	20, 37
NaI	^{23}Na, ^{127}I	37
$NaNO_3$	^{23}Na	20, 38
NaH	^1H	24
$NaHF_2$	^1H, ^{19}F	79*
$NaBH_4$	^1H	24
$Na_2B_4O_7.10H_2O$	^1H	39†
$Na_2SO_4.10H_2O$	^1H	39
$Na_2S_2O_3.5H_2O$	^{23}Na	87
$MgSO_4.7H_2O$	^1H	39
Al_2O_3	^{27}Al	20, 38
Si_2H_6	^1H	28†
P (red)	^{31}P	37
PH_4Br	^1H	82*
PH_4I	^1H	82*
KHF_2	^1H, ^{19}F	78, 79*
$KF.2H_2O$	^1H	39
KBr	^{79}Br, ^{81}Br	40
KI	^{127}I	20, 40
K_2BeF_4	^{19}F	25
KH_2PO_4	^1H	34†
KH_2AsO_4	^1H	34†
$KAl(SO_4)_2.12H_2O$	^1H	16, 39, 41
$KFe(SO_4)_2.12H_2O$	^1H	2

Solid	Resonant nuclei	References
$KCr(SO_4)_2.12H_2O$	^1H	16, 41, 49
$K_2Cu(SO_4)_2.6H_2O$	^1H	41
$K_2Mg(SO_4)_2.6H_2O$	^1H	41
$K_2Fe(SO_4)_2.6H_2O$	^1H	41
$K_3Cu(CN)_4$	$^{63}Cu, ^{65}Cu$	42
$K_2CuCl_4.2H_2O$	^1H	35*
$K_2HgCl_4.H_2O$	^1H	43*, 44*
$K_2SnCl_4.H_2O$	^1H	43*, 44*
$KB_5O_8.4H_2O$	^1H	13*
CaH_2	^1H	24
CaF_2	^{19}F	3, 4, 9, 16, 25, 30, 45, 46
$CaSO_4.2H_2O$	^1H	39*
V_2O_5	^{51}V	47
Cu_2Cl_2	$^{63}Cu, ^{65}Cu$	48
$CuCl_2.2H_2O$	^1H	35*, 49, 50*, 51, 77
$CuSO_4.5H_2O$	^1H	16, 49, 52
$ZnSO_4.7H_2O$	^1H	16
RbF, Cl, Br, I	^{87}Rb	53
Rb_2BeF_4	^{19}F	25
CsF, Cl, Br, I	^{133}Cs	53
$CsAl(SO_4)_2.12H_2O$	^1H	16
$Pb(VO_3)_2$	^{51}V	47
Glass	^{29}Si	27, 54, 55

Organic

Solid	Formula	Resonant nuclei	References
Methane	CH_4	^1H	7†, 8†, 56†
Monodeuteromethane	CH_3D	^1H	6†, 7†
n-Pentane	n-C_5H_{12}	^1H	57†
2.2-Dimethylpropane (neopentane)	C_5H_{12}	^1H	29†, 58†
n-Hexane	n-C_6H_{14}	^1H	57†
2.2-Dimethylbutane (neohexane)	C_6H_{14}	^1H	85†
Octadecane	n-$C_{18}H_{38}$	^1H	59†, 60†
Octacosane	n-$C_{28}H_{58}$	^1H	59†, 60†
Dicetyl	n-$C_{32}H_{66}$	^1H	59†, 60†
Paraffin wax		^1H	2, 45, 61, 62
Ethylene	C_2H_4	^1H	28
Benzene	C_6H_6	^1H	59†, 60†, 63*†, 64*†
Monodeuterobenzene	C_6H_5D	^1H	64*†
1.3.5-Trideuterobenzene	$C_6H_3D_3$	^1H	63*†, 64*†
o-Xylene	$C_6H_4(CH_3)_2$	^1H	59†, 60†
m-Xylene	$C_6H_4(CH_3)_2$	^1H	59†, 60†

Solid	Formula	Resonant nuclei	References
p-Xylene	$C_6H_4(CH_3)_2$	[1]H	59†, 60†
Mesitylene	$C_6H_3(CH_3)_3$	[1]H	59†, 60†
1.2.4.5-Tetramethyl-benzene (durene)	$C_6H_2(CH_3)_4$	[1]H	60†
Pentamethylbenzene	$C_6H(CH_3)_5$	[1]H	60†
Hexamethylbenzene	$C_6(CH_3)_6$	[1]H	59†, 60†
Hexaethylbenzene	$C_6(C_2H_5)_6$	[1]H	60†
Naphthalene	$C_{10}H_8$	[1]H	59
Anthracene	$C_{14}H_{10}$	[1]H	59, 65
Cyclopentane	C_5H_{10}	[1]H	57†, 85†
Cyclohexane	C_6H_{12}	[1]H	66*†
Styrene	C_8H_8	[1]H	75
Methyl iodide	CH_3I	[1]H	29†
Ethyl chloride	C_2H_5Cl	[1]H	29†
Ethyl bromide	C_2H_5Br	[1]H	29†
Ethyl iodide	C_2H_5I	[1]H	29†
1.2-Dichloroethane	$(CH_2Cl)_2$	[1]H	28*, 29†
1.1.1-Trichloroethane	$CH_3.CCl_3$	[1]H	28, 29†, 58†, 67
Hexafluoroethane	C_2F_6	[19]F	28, 29†
2.2-Dichloropropane	$(CH_3)_2CCl_2$	[1]H	58†
tert.-Butyl chloride	$(CH_3)_3CCl$	[1]H	58†, 68†
tert.-Butyl bromide	$(CH_3)_3CBr$	[1]H	58†
Methyl alcohol	CH_3OH	[1]H	29, 32†
Ethyl alcohol	C_2H_5OH	[1]H	29†
Lauryl alcohol	$n\text{-}C_{12}H_{25}OH$	[1]H	59†
Cetyl alcohol	$n\text{-}C_{16}H_{33}OH$	[1]H	69†
Ethylene glycol	$(CH_2OH)_2$	[1]H	65†
Acetone	$CH_3.CO.CH_3$	[1]H	29†
Oxalic acid dihydrate	$(COOH)_2.2H_2O$	[1]H	12*, 70*
Methylamine	CH_3NH_2	[1]H	29†
Urea	$OC(NH_2)_2$	[1]H	71*
Glycine	$CH_2.NH_2.COOH$	[1]H	72†
Acetonitrile	CH_3CN	[1]H	29†, 67
Nitromethane	CH_3NO_2	[1]H	29†
2.2-Dinitropropane	$C(CH_3)_2(NO_2)_2$	[1]H	58†
2-Nitro-2-chloropropane	$C(CH_3)_2NO_2Cl$	[1]H	58†
Dimethylsulphate	$(CH_3)_2SO_4$	[1]H	32†
Dimethylmercury	$(CH_3)_2Hg$	[1]H	29†
p-Azoxyanisole (liquid crystal)	$C_{14}H_{14}O_3N$	[1]H	73†
Polyethylene	$(CH_2)_n$	[1]H	74†, 88*
Polytetrafluoroethylene (Teflon)	$(CF_2)_n$	[19]F	30, 88*
Polymethyl-methacrylate (Plexiglass)	$(C_5H_8O_2)_n$	[1]H	75†
Polystyrene	$(CH_2.CH.C_6H_5)_n$	[1]H	75†
Rubber		[1]H	6†, 7†, 75†, 76†, 80†

1. Rollin and Hatton (1947). 2. Rollin, Hatton, Cooke and Benzie (1947). 3. Rollin and Hatton (1948). 4. Hatton and Rollin (1949). 5. Reif and Purcell (1953). 6. Alpert (1947). 7. Alpert (1949). 8. Bitter, Alpert, Poss, Lehr and Lin (1947). 9. Bloembergen, Purcell and Pound (1948). 10. Pake and Gutowsky (1948). 11. Sachs and Turner (1951). 12. Richards and Smith (1951). 13. Smith and Richards (1952). 14. Kakiuchi, Shono, Komatsu and Kigoshi (1951, 1952). 15. Kakiuchi and Komatsu (1952). 16. Bloembergen (1949 a). 17. Pound (1951). 18. Purcell and Pound (1951). 19. Ramsey and Pound (1951). 20. Pound (1950). 21. Soutif, Dreyfus and Ayant (1951); Soutif and Ayant (1953). 22. Schuster and Pake (1951 a). 23. Petch, Smellie and Volkoff (1951); Petch, Volkoff and Cranna (1952); Volkoff, Petch and Smellie (1952); Petch, Cranna and Volkoff (1953). 24. Garstens (1950). 25. Gutowsky, McClure and Hoffman (1951). 26. Schuster and Pake (1951 b). 27. Hatton, Rollin and Seymour (1951). 28. Gutowsky, Kistiakowsky, Pake and Purcell (1949). 29. Gutowsky and Pake (1950). 30. Pake and Purcell (1948). 31. Ayant and Soutif (1951). 32. Cooke and Drain (1952). 33. Gutowsky and Pake (1948); Gutowsky, Pake and Bersohn (1954). 34. Newman (1950 a). 35. Itoh, Kusaka, Yamagata, Kiriyama and Ibamoto (1953 a). 36. Pratt and Richards (1953); Deeley, Lewis and Richards (1954). 37. Pound (1948 b). 38. Taylor (1953). 39. Pake (1948). 40. Watkins and Pound (1953). 41. Darby and Rollin (1949). 42. Becker and Krüger (1951); Becker (1951). 43. Itoh, Kusaka, Yamagata, Kiriyama and Ibamoto (1952, 1953 b). 44. Itoh, Kusaka, Yamagata, Kiriyama, Ibamoto, Kanda and Masuda (1953). 45. Rollin (1946). 46. Purcell, Bloembergen and Pound (1946). 47. Knight and Cohen (1949). 48. Pound (1948 a). 49. Poulis (1951). 50. Poulis and Hardeman (1952 a, b, 1953 a, b, 1954); Poulis, Hardeman and Bölger (1952). 51. Poulis, van den Handel, Ubbink, Poulis and Gorter (1951). 52. Bloembergen (1950). 53. Gutowsky and McGarvey (1953 b). 54. Dharmatti and Weaver (1951 c). 55. Weaver (1953). 56. Thomas, Alpert and Torrey (1950); Tomita (1952, 1953). 57. Rushworth (1954). 58. Powles and Gutowsky (1953 a). 59. Andrew (1950). 60. Andrew (1951 a). 61. Purcell, Torrey and Pound (1946). 62. Bloch, Hansen and Packard

(1946 *a*, *b*).   63. Andrew and Eades (1953 *a*).   64. Andrew and Eades (1953 *c*).   65. Rushworth (1952).   66. Andrew and Eades (1952, 1953 *b*).   67. Andrew and Bersohn (1950).   68. Powles and Gutowsky (1953 *b*).   69. Kojima, Ogawa and Torizuka (1951); Kojima and Ogawa (1953).   70. Itoh, Kusaka, Kiriyama and Yabumoto (1953).   71. Andrew and Hyndman (1953).   72. Shaw, Elsken and Palmer (1952).   73. Spence, Moses and Jain (1953); Spence, Gutowsky and Holm (1953).   74. Newman (1950 *b*). 75. Holroyd, Codrington, Mrowca and Guth (1951).   76. Cohen-Hadria and Gabillard (1952).   77. Gorter (1953).   78. Van Vleck (1948).   79. Waugh, Humphrey and Yost (1953).   80. Gutowsky and Meyer (1953).   81. Deeley and Richards (1954).   82. Pratt and Richards (1954).   83. Bersohn and Gutowsky (1954).   84. Soutif (1951).   85. Aston, Bolger, Trambarulo and Segall (1954). 86. Tomita and Mannari (1953).   87. Itoh, Kusaka and Yamagata (1954).   88. Wilson and Pake (1953).

METALS INVESTIGATED BY THE NUCLEAR MAGNETIC RESONANCE METHOD

Metal	Isotope	References
Lithium	^{7}Li	Knight (1949); Poulis (1950); Gutowsky (1951); Gutowsky and McGarvey (1952)
Beryllium	^{9}Be	Townes, Herring and Knight (1950); Knight (1953)
Sodium	^{23}Na	Knight (1949); Gutowsky (1951); Norberg and Slichter (1951); Gutowsky and McGarvey (1952)
Aluminium	^{27}Al	Rollin and Hatton (1948); Knight (1949); Hatton and Rollin (1949); Poulis (1950); Gutowsky (1951); Gutowsky and McGarvey (1952); Seymour (1953)
Silicon	^{29}Si	Dharmatti and Weaver (1951c)
Vanadium	^{51}V	Knight (1952); Walchli and Morgan (1952)
Copper	$^{63,\ 65}$Cu	Pound (1948b); Knight (1949); Bloembergen (1949b); Hatton and Rollin (1949); Gutowsky (1951); Gutowsky and McGarvey (1952); Bloembergen (1952)
Gallium	$^{69,\ 71}$Ga	Knight (1949); Townes, Herring and Knight (1950); Gutowsky and McGarvey (1952); McGarvey and Gutowsky (1953)
Rubidium	$^{85,\ 87}$Rb	Gutowsky and McGarvey (1952); McGarvey and Gutowsky (1953)
Niobium	^{93}Nb	Knight (1952)
Silver	$^{107,\ 109}$Ag	Brun, Oeser, Staub and Telschow (1954a); Sogo and Jeffries (1954)
Tin	$^{117,\ 119}$Sn	Bloembergen and Rowland (1953); McGarvey and Gutowsky (1953)
Tellurium	$^{123,\ 125}$Te	Dharmatti and Weaver (1951c)
Caesium	^{133}Cs	Gutowsky and McGarvey (1952); McGarvey and Gutowsky (1953)
Thallium	$^{203,\ 205}$Tl	Bloembergen and Rowland (1953)
Lead	^{207}Pb	Townes, Herring and Knight (1950); Bloembergen and Rowland (1953)

SPECTRUM FOR SYSTEMS OF TWO NUCLEI OF SPIN NUMBER ½†

Let us consider a crystal containing relatively isolated pairs of identical nuclei of spin number ½ rigidly held to their lattice sites. A good example is provided by gypsum, $CaSO_4 . 2H_2O$, which was discussed in §6.2.1, and which contains well-separated pairs of protons. As a first approximation we may regard the crystal as a number of independent nuclear magnetic pairs, each nucleus interacting with its partner only. The magnetic energy levels are perturbed by the dipole-dipole interaction, which from equation (2) of Appendix 3 is

$$V_{12} = \gamma^2 \hbar^2 r^{-3} [\mathbf{I}_1 . \mathbf{I}_2 - 3(\mathbf{I}_1 . \mathbf{p})(\mathbf{I}_2 . \mathbf{p})], \tag{1}$$

where $\mathbf{p}$ is a unit vector directed from one nucleus to the other, r is their separation, $\mathbf{I}\hbar$ is the nuclear spin operator and γ is the nuclear gyromagnetic ratio $\mu/I\hbar$, where μ is the nuclear magnetic moment.

Since the lattice co-ordinates of the nuclei are considered to remain constant, the spin wave functions for the perturbation problem include both symmetric and antisymmetric representations. We choose therefore singlet and triplet functions for the two-spin system. Transitions between the triplet levels and the singlet level are highly forbidden, and we need only consider the diagonal elements $\langle m \,|\, V_{12} \,|\, m \rangle$ for the triplet levels, where m, the magnetic quantum number for the two-spin system, is $+1, 0, -1$ for the three triplet levels.

In Appendix 3 we saw (equation (7)) that V_{12} could be rewritten as

$$V_{12} = \gamma^2 \hbar^2 r^{-3} (A + B + C + D + E + F), \tag{2}$$

where the functions A to F are given in equations (8) of Appendix 3. Here, for two nuclei, we put $i = 1$, $j = 2$ and we can omit the

† Pake, (1948).

subscripts from the polar angle θ. The contribution of the term involving A to the diagonal elements is therefore

$$\langle m \mid (V_{12})_A \mid m \rangle = -\gamma^2\hbar^2 r^{-3}(3\cos^2\theta - 1)\langle m \mid I_{z_1}I_{z_2} \mid m \rangle.$$

Thus we have

$$\left.\begin{aligned}
\langle \pm 1 \mid (V_{12})_A \mid \pm 1 \rangle &= -\gamma^2\hbar^2 I^2 r^{-3}(3\cos^2\theta - 1) \\
&= -\mu^2 r^{-3}(3\cos^2\theta - 1), \\
\langle 0 \mid (V_{12})_A \mid 0 \rangle &= +\gamma^2\hbar^2 I^2 r^{-3}(3\cos^2\theta - 1) \\
&= +\mu^2 r^{-3}(3\cos^2\theta - 1).
\end{aligned}\right\} \quad (3)$$

The only other term in (2) which contributes is that involving B:

$$\begin{aligned}
(V_{12})_B &= \tfrac{1}{4}\gamma^2\hbar^2 r^{-3}(3\cos^2\theta - 1)[(I_{x_1} - iI_{y_1})(I_{x_2} + iI_{y_2}) \\
&\qquad\qquad + (I_{x_1} + iI_{y_1})(I_{x_2} - iI_{y_2})] \quad (4) \\
&= \tfrac{1}{2}\gamma^2\hbar^2 r^{-3}(3\cos^2\theta - 1)[I_{x_1}I_{x_2} + I_{y_1}I_{y_2}] \\
&= \tfrac{1}{2}\gamma^2\hbar^2 r^{-3}(3\cos^2\theta - 1)[I^2 - I_{z_1}I_{z_2}]. \quad (5)
\end{aligned}$$

Thus we have

$$\left.\begin{aligned}
\langle \pm 1 \mid (V_{12})_B \mid \pm 1 \rangle &= 0, \\
\langle 0 \mid (V_{12})_B \mid 0 \rangle &= \gamma^2\hbar^2 I^2 r^{-3}(3\cos^2\theta - 1) = \mu^2 r^{-3}(3\cos^2\theta - 1).
\end{aligned}\right\} (6)$$

Combining (3) and (6) we get for the diagonal elements

$$\left.\begin{aligned}
\langle \pm 1 \mid V_{12} \mid \pm 1 \rangle &= -\mu^2 r^{-3}(3\cos^2\theta - 1), \\
\langle 0 \mid V_{12} \mid 0 \rangle &= +2\mu^2 r^{-3}(3\cos^2\theta - 1).
\end{aligned}\right\} \quad (7)$$

The unperturbed and perturbed energy levels for the two-spin

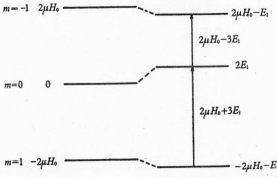

Fig. 61. Energy level diagram for a system of two identical nuclei showing the effect of the perturbing dipole-dipole interaction. In the diagram E_1 represents $\mu r^{-3}(3\cos^2\theta - 1)$.

system are shown in fig. 61. Resonance absorption occurs for the perturbed system when

$$h\nu = 2\mu H_0 \pm 3\mu r^{-3}(3\cos^2\theta - 1). \quad (8)$$

Substitution of $h\nu = 2\mu H^*$ yields the resonance value of the external field as

$$H_0 = H^* \pm \tfrac{3}{2}\mu r^{-3}(3\cos^2\theta - 1), \qquad (9)$$

which is the result quoted in §6.2.1, equation (6.2).

Let us now consider the case in which the isolated pairs consist of two different nuclei of spin number $\tfrac{1}{2}$, with gyromagnetic ratios γ_1, γ_2 and magnetic moments μ_1, μ_2. Expressions (1) and (2) for the interaction energy must be modified by replacing γ^2 with $\gamma_1\gamma_2$. The contribution of the term involving A to the diagonal elements now becomes

$$\left.\begin{aligned}\langle \pm 1 \mid (V_{12})_A \mid \pm 1\rangle &= -\mu_1\mu_2 r^{-3}(3\cos^2\theta - 1), \\ \langle 0 \mid (V_{12})_A \mid 0\rangle &= +\mu_1\mu_2 r^{-3}(3\cos^2\theta - 1).\end{aligned}\right\} \qquad (10)$$

This is, in fact, the only contribution for this case, since the term involving B no longer contributes on account of the different Larmor precessional frequencies of the two nuclei. The resonance condition for the nuclei with moment μ_1 is therefore

$$h\nu = 2\mu_1 H_0 \pm 2\mu_1\mu_2 r^{-3}(3\cos^2\theta - 1). \qquad (11)$$

Substitution of $h\nu = 2\mu_1 H^*$ yields the resonance value of the external field as

$$H_0 = H^* \pm \mu_2 r^{-3}(3\cos^2\theta - 1), \qquad (12)$$

in which the factor $\tfrac{3}{2}$ no longer appears (cf. equation (9)).

GLOSSARY OF SYMBOLS

The page numbers indicate where the symbols are introduced.

$\mathscr{I}$	electric current	33
$\mathscr{L}$	coil inductance	33
$\mathscr{M}$	magnetization	22
$\mathscr{N}$	number of interacting nuclei in Van Vleck's second moment formula	159
$\mathscr{Q}$	quality factor of tuned circuit	33
$\mathscr{R}$	resistance	35
$\mathscr{V}$	electric potential	33
$\mathscr{Y}$	correction factor in equation (3.26)	69
$\mathscr{Z}$	impedance	39

α	fine structure constant	95
β	Bohr magneton	93
γ	nuclear gyromagnetic ratio	9
γ_{jk}	angle used in §6.4	167
δ	small change, a general variable; in §5.5.1, magnetic field parameter	130
ϵ	angle, a general variable; in Appendix 3, direction cosine	230
ζ	filling factor	31
η	viscosity	116
θ	angle, a general variable	
κ	demagnetizing factor	78
λ	unit vector in Appendix 3	230
μ	magnetic moment, usually nuclear	3
μ_0	nuclear magneton	8
ν	frequency, a general variable; ν_0 resonance frequency	10
ξ	numerical factor in §5.3	119
ρ	β-ray orbit radius in §4.11; charge density in §8.1	102 ; 203
σ	saturation parameter	109
τ_c	correlation time	115
ϕ	angle, a general variable	
χ	magnetic susceptibility; χ', χ'' real and imaginary parts of susceptibility; χ_0 static susceptibility	21
ψ	wave function	197
ω	angular frequency ($=2\pi\nu$)	
Γ	wave function ratio	197
Δ	increment	

R

BIBLIOGRAPHY AND AUTHOR INDEX

The bibliography contains a fairly complete list of papers published before the middle of 1954. Conference abstracts are not included except when they contain important information not available elsewhere.

(1) *Some general references*

BLOCH, F. (1946). *Phys. Rev.* **70**, 460.

BLOCH, F. (1953). *Science*, **118**, 425; *Phys. Bl.* **9**, 442.

BLOCH, F., HANSEN, W. W. and PACKARD, M. E. (1946). *Phys. Rev.* **70**, 474.

BLOEMBERGEN, N., PURCELL, E. M. and POUND, R. V. (1948). *Phys. Rev.* **73**, 679.

BRAUNBEK, W. (1950). *Phys. Bl.* **6**, 5.

DARROW, K. K. (1953). *Bell Syst. tech. J.* **32**, 74.

GIULOTTO, L. (1953). *R. C. Semin. mat. fis. Milano*, **24**.

GUTOWSKY, H. S. (1954). *Ann. Rev. Phys. Chem.* **5**, 333.

PAKE, G. E. (1950). *Amer. J. Phys.* **18**, 438, 473.

POUND, R. V. (1952). *Progr. Nucl. Phys.* **2**, 21.

PURCELL, E. M. (1948). *Science*, **107**, 433.

PURCELL, E. M. (1951). *Physica*, **17**, 282.

PURCELL, E. M. (1953). *Science*, **118**, 431; *Phys. Bl.* **9**, 453.

PURCELL, E. M. (1954). *Amer. J. Phys.* **22**, 1.

RAMSEY, N. F. (1953). *Nuclear Moments*. New York: John Wiley and Sons, Inc.

ROBERTS, A. (1947). *Nucleonics*, **1**.

ROLLIN, B. V. (1949). *Rep. Progr. Phys.* **12**, 22.

SMITH, J. A. S. (1953). *Quart. Rev. Chem. Soc.* **7**, 279.

SOUTIF, M. (1949). *J. Phys. Radium*, (8), **10**, 61 D.

(2) *Original papers and books quoted in the text*

The figures in italics indicate the pages of the text in which the reference is quoted.

ADAMS, N. I., WIMETT, T. F. and BITTER, F. (1951). *Phys. Rev.* **82**, 343. [*228*]

ALDER, F. and HALBACH, K. (1953). *Helv. phys. acta*, **26**, 426. [*228*]

ALDER, F. and YU, F. C. (1951 a). *Phys. Rev.* **81**, 1067. [*228*]

ALDER, F. and YU, F. C. (1951 b). *Phys. Rev.* **82**, 105. [*228*]

ALPERT, N. L. (1947). *Phys. Rev.* **72**, 637. [*165, 239*]

ALPERT, N. L. (1949). *Phys. Rev.* **75**, 398. [*165, 239*]

ANDERSON, D. A. (1949). *Phys. Rev.* **76**, 434. [*228*]

ANDERSON, H. L. (1949). *Phys. Rev.* **76**, 1460. [*43, 47, 48, 71, 76, 149, 228*]

ANDERSON, H. L. and NOVICK, A. (1947). *Phys. Rev.* **71**, 372. [*228*]

ANDERSON, H. L. and NOVICK, A. (1948). *Phys. Rev.* **73**, 919. [*228*]

ANDERSON, W. A. and ARNOLD, J. T. (1954). *Phys. Rev.* **94**, 497. [*146*]

ANDREW, E. R. (1950). *J. Chem. Phys.* **18**, 607. [*167, 171, 239*]

ANDREW, E. R. (1951 a). *Physica*, **17**, 405. [*239*]

ANDREW, E. R. (1951 b). *Phys. Rev.* **82**, 443. [*144*]

ANDREW, E. R. (1953). *Phys. Rev.* **91**, 425. [*46, 164*]

ANDREW, E. R. and BERSOHN, R. (1950). *J. Chem. Phys.* **18**, 159; **20**, 924. [*156, 169, 240*]

ANDREW, E. R. and EADES, R. G. (1952). *Proc. Phys. Soc. Lond.* A, **65**, 371. [*172, 240*]

ANDREW, E. R. and EADES, R. G. (1953*a*). *Proc. Phys. Soc. Lond.* A, **66**, 415. [*163, 171, 240*]

ANDREW, E. R. and EADES, R. G. (1953*b*). *Proc. Roy. Soc.* A, **216**, 398. [*163, 171, 172, 173, 179, 240*]

ANDREW, E. R. and EADES, R. G. (1953*c*). *Proc. Roy. Soc.* A, **218**, 537. [*163, 171, 173, 178, 240*]

ANDREW, E. R. and HYNDMAN, D. (1953). *Proc. Phys. Soc. Lond.* A, **66**, 1187. [*162, 240*]

ANDREW, E. R. and RUSHWORTH, F. A. (1952). *Proc. Phys. Soc. Lond.* B, **65**, 801. [*73*]

ARNOLD, J. T., DHARMATTI, S. S. and PACKARD, M. E. (1951). *J. Chem. Phys.* **19**, 507. [*142, 143*]

ARNOLD, J. T. and PACKARD, M. E. (1951). *J. Chem. Phys.* **19**, 1608. [*142*]

ARNOLD, W. R. and ROBERTS, A. (1946). *Phys. Rev.* **70**, 766. [*227*]

ARNOLD, W. R. and ROBERTS, A. (1947). *Phys. Rev.* **71**, 878. [*227*]

ASTON, J. G., BOLGER, B., TRAMBARULO, R. and SEGALL, H. (1954). *J. Chem. Phys.* **22**, 460. [*240*]

AYANT, Y. (1951*a*). *C.R. Acad. Sci., Paris*, **232**, 1203. [*79*]

AYANT, Y. (1951*b*). *C.R. Acad. Sci., Paris*, **232**, 1298. [*79*]

AYANT, Y. (1951*c*). *C.R. Acad. Sci., Paris*, **233**, 245. [*136*]

AYANT, Y. (1953). *C.R. Acad. Sci., Paris*, **236**, 198. [*119*]

AYANT, Y. (1954). *C.R. Acad. Sci., Paris*, **238**, 1876. [*205*]

AYANT, Y. and SOUTIF, M. (1951). *C.R. Acad. Sci., Paris*, **232**, 639. [*239*]

BAKER, E. B. (1954). *Rev. Sci. Instrum.* **25**, 390. [*59*]

BAYER, H. (1951). *Z. Phys.* **130**, 227. [*219*]

BECKER, G. (1951). *Z. Phys.* **130**, 415. [*87, 239*]

BECKER, G. and KRÜGER, H. (1951). *Naturwissenschaften*, **38**, 12. [*87, 239*]

BÉNÉ, G. J. (1951). *Helv. phys. acta*, **24**, 367. [*100, 228*]

BÉNÉ, G. J., DENIS, P. M. and EXTERMANN, R. C. (1950*a*). *Arch. Sci., Genève*, **3**, 49. [*99*]

BÉNÉ, G. J., DENIS, P. M. and EXTERMANN, R. C. (1950*b*). *Phys. Rev.* **77**, 288. [*100*]

BÉNÉ, G. J., DENIS, P. M. and EXTERMANN, R. C. (1950*c*). *Arch. Sci., Genève*, **3**, 452. [*136*]

BÉNÉ, G. J., DENIS, P. M. and EXTERMANN, R. C. (1950*d*). *C.R. Acad. Sci., Paris*, **231**, 1294. [*136*]

BÉNÉ, G. J., DENIS, P. M. and EXTERMANN, R. C. (1951*a*). *Helv. phys. acta*, **24**, 633. [*99*]

BÉNÉ, G. J., DENIS, P. M. and EXTERMANN, R. C. (1951*b*). *Helv. phys. acta*, **24**, 304. [*100, 136*]

BÉNÉ, G. J., DENIS, P. M. and EXTERMANN, R. C. (1951*c*). *Arch. Sci., Genève*, **4**, 212. [*136*]

BÉNÉ, G. J., DENIS, P. M. and EXTERMANN, R. C. (1951*d*). *Arch. Sci., Genève*, **4**, 266. [*136*]

BÉNÉ, G. J., DENIS, P. M. and EXTERMANN, R. C. (1951 e). *Physica,* **17,** 308. [*135*]

BÉNÉ, G. J., DENIS, P. M. and EXTERMANN, R. C. (1952 a). *J. Phys. Radium,* **13,** 71 S. [*99*]

BÉNÉ, G. J., DENIS, P. M. and EXTERMANN, R. C. (1952 b). *Arch. Sci., Genève,* **5,** 406. [*136*]

BÉNÉ, G. J., DENIS, P. M. and EXTERMANN, R. C. (1953 a). *Helv. phys. acta,* **26,** 267. [*136*]

BÉNÉ, G. J., DENIS, P. M., EXTERMANN, R. C. and BONHOMME, H. J. (1953 b). *Helv. phys. acta,* **26,** 435. [*136*]

BERSOHN, R. (1952). *J. Chem. Phys.* **20,** 1505. [*210, 220*]

BERSOHN, R. and GUTOWSKY, H. S. (1954). *J. Chem. Phys.* **22,** 651. [*158, 162, 238*]

BHAR, J. N. and BHAR, B. N. (1952). *Sci. and Culture,* **18,** 86. [*136*]

BITTER, F. (1949 a). *Phys. Rev.* **75,** 1326. [*228*]

BITTER, F. (1949 b). *Phys. Rev.* **76,** 150. [*75*]

BITTER, F., ALPERT, D. N., NAGLE, D. E. and POSS, H. L. (1947). *Phys. Rev.* **72,** 1271. [*227*]

BITTER, F., ALPERT, D. N., POSS, H. L., LEHR, C. G. and LIN, S. T. (1947). *Phys. Rev.* **71,** 738. [*239*]

BITTER, F. and BROSSEL, J. (1952). *Phys. Rev.* **85,** 1051. [*65*]

BITTER, F., LACEY, R. F. and RICHTER, B. (1953). *Rev. Mod. Phys.* **25,** 174. [*65*]

BLEANEY, B., DANIELS, J. M., GRACE, M. A., HALBAN, H., KURTI, N., ROBINSON, F. N. H. and SIMON, F. E. (1954). *Proc. Roy. Soc.* A, **221,** 170. [*66*]

BLOCH, F. (1946). *Phys. Rev.* **70,** 460. [*15, 17, 18, 25, 130, 131, 222*]

BLOCH, F. (1951 a). *Physica,* **17,** 272. [*33*]

BLOCH, F. (1951 b). *Phys. Rev.* **83,** 1062. [*149*]

BLOCH, F. (1954). *Phys. Rev.* **94,** 496. [*145*]

BLOCH, F. and GARBER, D. H. (1949). *Phys. Rev.* **76,** 585. [*46*]

BLOCH, F., GRAVES, A. C., PACKARD, M. E. and SPENCE, R. W. (1947 a). *Phys. Rev.* **71,** 373. [*228*]

BLOCH, F., GRAVES, A. C., PACKARD, M. E. and SPENCE, R. W. (1947 b). *Phys. Rev.* **71,** 551. [*228*]

BLOCH, F., HANSEN, W. W. and PACKARD, M. E. (1946 a). *Phys. Rev.* **69,** 127. [*5, 34, 56, 239*]

BLOCH, F., HANSEN, W. W. and PACKARD, M. E. (1946 b). *Phys. Rev.* **70,** 474. [*34, 56, 58, 126, 130, 131, 239*]

BLOCH, F. and JEFFRIES, C. D. (1950). *Phys. Rev.* **80,** 305. [*91, 92, 227*]

BLOCH, F., LEVINTHAL, E. C. and PACKARD, M. E. (1947). *Phys. Rev.* **72,** 1825. [*227*]

BLOCH, F., NICODEMUS, D. and STAUB, H. H. (1948). *Phys. Rev.* **74,** 1025. [*227*]

BLOCH, F. and SIEGERT, A. (1940). *Phys. Rev.* **57,** 522. [*80*]

BLOEMBERGEN, N. (1948). *Nuclear Magnetic Relaxation.* The Hague: Nijhoff. [*5, 40, 41, 45, 105, 110, 111, 132, 204, 205, 228, 233*]

BLOEMBERGEN, N. (1949 a). *Physica,* **15,** 386. [*176, 177, 239*]

BLOEMBERGEN, N. (1949 b). *Physica,* **15,** 588. [*193, 194, 195, 241*]

BLOEMBERGEN, N. (1950). *Physica*, **16**, 95.   [*185, 186, 239*]
BLOEMBERGEN, N. (1952). *J. Appl. Phys.* **23**, 1383.   [*191, 241*]
BLOEMBERGEN, N. and DICKINSON, W. C. (1950). *Phys. Rev.* **79**, 179.   [*79*]
BLOEMBERGEN, N. and POULIS, N. J. (1950). *Physica*, **16**, 915.   [*187*]
BLOEMBERGEN, N., PURCELL, E. M. and POUND, R. V. (1947).
    Nature, Lond. **160**, 475.   [*105*]
BLOEMBERGEN, N., PURCELL, E. M. and POUND, R. V. (1948).
    Phys. Rev. **73**, 679. [*7, 18, 19, 25, 37, 40, 41, 47, 70, 105, 116,
                    117, 121–8, 132, 150, 205, 230, 239*]
BLOEMBERGEN, N. and ROWLAND, T. J. (1953). *Acta Metallurgica*,
    1, 731.   [*199, 201, 211, 241*]
BLOEMBERGEN, N. and TEMMER, G. M. (1953). *Phys. Rev.* **89**, 883.   [*66*]
BLOOM, A. L. and SHOOLERY, J. N. (1953). *Phys. Rev.* **90**, 358.   [*112*]
BOHR, A. and WEISSKOPF, V. F. (1950). *Phys. Rev.* **77**, 94.   [*75*]
BOLLE, A., PUPPI, G. and ZANOTELLI, G. (1946). *Nuovo Cim.* **3**, 412. [*228*]
BOLLE, A. and ZANOTELLI, G. (1948a). *Ric. sci.* **18**, 847.   [*228*]
BOLLE, A. and ZANOTELLI, G. (1948b). *R. C. Accad. XL*, **27**.   [*136*]
BRADFORD, R., CLAY, C., CRAFT, A., STRICK, E. and UNDERHILL, J.
    (1951a). *Phys. Rev.* **83**, 656.   [*141*]
BRADFORD, R., CLAY, C. and STRICK, E. (1951b). *Phys. Rev.* **84**, 157. [*141*]
BROER, L. F. J. (1945). Thesis, Amsterdam.   [*15*]
BROSSEL, J., CAGNAC, B. and KASTLER, A. (1954). *J. Phys. Radium*,
    15, 6.   [*65*]
BROSSEL, J., KASTLER, A. and WINTER, J. (1952). *J. Phys. Radium*,
    13, 668.   [*65*]
BROWN, R. M. (1950). *Phys. Rev.* **78**, 530.   [*47, 49, 146*]
BROWN, R. M. and PURCELL, E. M. (1949). *Phys. Rev.* **75**, 1262.   [*146*]
BROWN, W. L. (1951). *Phys. Rev.* **83**, 271.   [*102*]
BRUN, E., OESER, J., STAUB, H. H. and TELSCHOW, C. G. (1954a).
    Phys. Rev. **93**, 172.   [*75, 228, 241*]
BRUN, E., OESER, J., STAUB, H. H. and TELSCHOW, C. G. (1954b).
    Phys. Rev. **93**, 904.   [*148, 229*]
BURGESS, J. H. and BROWN, R. M. (1952). *Rev. Sci. Instrum.* **23**, 334. [*146*]
CARR, E. F. and KIKUCHI, C. (1950). *Phys. Rev.* **78**, 470.   [*210*]
CARR, H. Y. and PURCELL, E. M. (1952). *Phys. Rev.* **88**, 415.   [*145*]
CARR, H. Y. and PURCELL, E. M. (1954). *Phys. Rev.* **94**, 630.   [*141*]
CARVER, T. R. and SLICHTER, C. P. (1953). *Phys. Rev.* **92**, 212.   [*196*]
CHAMBERS, W. H. and WILLIAMS, D. (1949). *Phys. Rev.* **76**, 638.   [*228*]
CHANDRASEKHAR, S. (1943). *Rev. Mod. Phys.* **15**, 1.   [*117*]
CHIAROTTI, G. and GIULOTTO, L. (1951). *Nuovo Cim.* **8**, 595.   [*228*]
CHIAROTTI, G. and GIULOTTO, L. (1953). *Nuovo Cim.* **10**, 54.   [*127*]
CHIAROTTI, G. and GIULOTTO, L. (1954). *Phys. Rev.* **93**, 1241.   [*117*]
CLAY, C. S., BRADFORD, R. S. and STRICK, E. (1951). *J. Chem. Phys.*
    19, 1429.   [*141*]
COHEN, V. W., KNIGHT, W. D., WENTINK, T. and KOSKI, W. S.
    (1950). *Phys. Rev.* **79**, 191.   [*228*]
COHEN-HADRIA, A. and GABILLARD, R. (1952). *C.R. Acad. Sci.,
    Paris*, **234**, 1877.   [*240*]
COLLINS, T. L. (1950a). *Phys. Rev.* **79**, 226.   [*228*]

COLLINS, T. L. (1950 b). *Phys. Rev.* **80**, 103. [228]

CONDON, E. U. and SHORTLEY, G. H. (1935). *The Theory of Atomic Spectra*. Cambridge University Press. [19, 233]

CONGER, R. L. (1953). *J. Chem. Phys.* **21**, 937. [128]

CONGER, R. L. and SELWOOD, P. W. (1952). *J. Chem. Phys.* **20**, 383. [128]

COOKE, A. H. and DRAIN, L. E. (1952). *Proc. Phys. Soc. Lond.* A, **65**, 894. [175, 177, 239]

COX, H. L. (1953). *Rev. Sci. Instrum.* **24**, 307. [45]

CRANNA, N. G. (1953). *Canad. J. Phys.* **31**, 1185. [87]

CRAWFORD, M. F. and SCHAWLOW, A. L. (1949). *Phys. Rev.* **76**, 1310. [75]

DANIELS, J. M., GRACE, M. A., HALBAN, H., KURTI, N. and ROBINSON, F. N. H. (1952). *Phil. Mag.* **43**, 1297. [66]

DANIELS, J. M., GRACE, M. A. and ROBINSON, F. N. H. (1951). *Nature, Lond.* **168**, 780. [66]

DARBY, J. F. and ROLLIN, B. V. (1949). *Nature, Lond.* **164**, 66. [239]

DAS, T. P. and SAHA, A. K. (1954). *Phys. Rev.* **93**, 749. [139]

DEBYE, P. (1945). *Polar Molecules*. New York: Dover. [78, 116]

DEELEY, C. M., LEWIS, P. and RICHARDS, R. E. (1954). *Trans. Faraday Soc.* **50**, 556. [239]

DEELEY, C. M. and RICHARDS, R. E. (1954). *Trans. Faraday Soc.* **50**, 560. [240]

DEHMELT, H. G. (1951). *Z. Phys.* **130**, 356. [218, 220]

DEHMELT, H. G. (1954). *Amer. J. Phys.* **22**, 110. [216, 217, 220]

DEHMELT, H. G. and KRÜGER, H. (1950). *Naturwissenschaften*, **37**, 111. [87, 216]

DENIS, P. M., BÉNÉ, G. J. and EXTERMANN, R. C. (1952). *Arch. Sci., Genève*, **5**, 32. [112, 136]

DHARMATTI, S. S. and WEAVER, H. E. (1951 a). *Phys. Rev.* **83**, 845. [228]

DHARMATTI, S. S. and WEAVER, H. E. (1951 b). *Phys. Rev.* **84**, 367. [228]

DHARMATTI, S. S. and WEAVER, H. E. (1951 c). *Phys. Rev.* **84**, 843. [228, 239, 241]

DHARMATTI, S. S. and WEAVER, H. E. (1952 a). *Phys. Rev.* **85**, 927. [228]

DHARMATTI, S. S. and WEAVER, H. E. (1952 b). *Phys. Rev.* **86**, 259. [228]

DHARMATTI, S. S. and WEAVER, H. E. (1952 c). *Phys. Rev.* **87**, 675. [143]

DICKE, R. H. (1946). *Rev. Sci. Instrum.* **17**, 268. [45]

DICKINSON, W. C. (1949). *Phys. Rev.* **76**, 1414. [228]

DICKINSON, W. C. (1950 a). *Phys. Rev.* **77**, 736. [141]

DICKINSON, W. C. (1950 b). *Phys. Rev.* **80**, 563. [80]

DICKINSON, W. C. (1951). *Phys. Rev.* **81**, 717. [78, 79, 141, 142]

DICKINSON, W. C. and WIMETT, T. F. (1949). *Phys. Rev.* **75**, 1769. [228]

DRAIN, L. E. (1949). *Proc. Phys. Soc. Lond.* A, **62**, 301. [132]

DUMOND, J. W. M. and COHEN, E. R. (1953). *Rev. Mod. Phys.* **25**, 691. [96]

EINSTEIN, A. (1917). *Phys. Z.* **18**, 121. [11]

ESTERMANN, I. and STERN, O. (1933). *Z. Phys.* **85**, 17. [3]

EXTERMANN, R. C., DENIS, P. M. and BÉNÉ, G. J. (1949). *Helv. phys. acta*, **22**, 388. [136]

FAIRBANK, W. M., ARD, W. B., DEHMELT, H. G., GORDY, W. and WILLIAMS, S. R. (1953). *Phys. Rev.* **92**, 208. [150]

FAIRBANK, W. M., ARD, W. B. and WALTERS, G. K. (1954). *Phys. Rev.* **95**, 566.   [*150*]

FERMI, E. (1930). *Z. Phys.* **60**, 320.   [*197*]

FRISCH, R. and STERN, O. (1933). *Z. Phys.* **85**, 4.   [*3*]

GABILLARD, R. (1951*a*). *C.R. Acad. Sci., Paris*, **232**, 324.   [*35*]

GABILLARD, R. (1951*b*). *C.R. Acad. Sci., Paris*, **232**, 1477.   [*129*]

GABILLARD, R. (1951*c*). *C.R. Acad. Sci., Paris*, **232**, 1551.   [*134, 135*]

GABILLARD, R. (1951*d*). *C.R. Acad. Sci., Paris*, **233**, 39.   [*135*]

GABILLARD, R. (1951*e*). *C.R. Acad. Sci., Paris*, **233**, 307.   [*136*]

GABILLARD, R. (1952*a*). *Phys. Rev.* **85**, 694.   [*134,135*]

GABILLARD, R. (1952*b*). *Rev. Sci., Paris*, **90**, 307.   [*127, 128, 134, 135*]

GABILLARD, R. and SOUTIF, M. (1950). *C.R. Acad. Sci., Paris*, **230**, 1754.   [*53*]

GABILLARD, R. and SOUTIF, M. (1951). *C. R. Acad. Sci., Paris*, **233**, 480.   [*135*]

GARDNER, J. H. (1951). *Phys. Rev.* **83**, 996,   [*93, 94, 227*]

GARDNER, J. H. and PURCELL, E. M. (1949). *Phys. Rev.* **76**, 1262.
                                                             [*93, 94, 227*]

GARSTENS, M. A. (1950). *Phys. Rev.* **79**, 397.   [*149, 239*]

GARSTENS, M. A. (1951). *Phys. Rev.* **81**, 288.   [*149*]

GERLACH, W. and STERN, O. (1924). *Ann. Phys., Lpz.* **74**, 673.   [*2*]

GINDSBERG, J. and BEERS, Y. (1953). *Rev. Sci. Instrum.* **24**, 632.   [*53*]

GIULOTTO, L. (1948). *Nuovo Cim.* **5**, 498.   [*136*]

GIULOTTO, L. and GIGLI, A. (1947). *Nuovo Cim.* **4**, 275.   [*60*]

GIULOTTO, L., GIGLI, A. and SILLANO, P. (1947). *Nuovo Cim.* **4**, 201.   [*60*]

GOODEN, J. S. (1950). *Nature, Lond.* **165**, 1014.   [*135*]

GORTER, C. J. (1936). *Physica*, **3**, 995.   [*4*]

GORTER, C. J. (1951). *Physica*, **17**, 169.   [*4*]

GORTER, C. J. (1953). *Rev. Mod. Phys.* **25**, 332.   [*189, 240*]

GORTER, C. J. and BROER, L. F. J. (1942). *Physica*, **9**, 591.   [*4*]

GORTER, C. J. and KRONIG, R. de L. (1936). *Physica*, **3**, 1009.   [*23*]

GORTER, C. J., POPPEMA, O. J., STEENLAND, M. J. and BEUN, J. A. (1951). *Physica*, **17**, 1050.   [*66*]

GRISWOLD, T. W., KIP, A. F. and KITTEL, C. (1952). *Phys. Rev.* **88**, 951.   [*196*]

GRIVET, P. (1951). *C.R. Acad. Sci., Paris*, **233**, 397.   [*69*]

GRIVET, P. and AYANT, Y. (1951). *C.R. Acad. Sci., Paris*, **232**, 1094.   [*79*]

GRIVET, P. and SOUTIF, M. (1949). *C.R. Acad. Sci., Paris*, **228**, 1852. [*55*]

GRIVET, P., SOUTIF, M. and BUYLE, M. (1949). *C.R. Acad. Sci., Paris*, **229**, 113.   [*47, 48*]

GRIVET, P., SOUTIF, M. and GABILLARD, R. (1949). *C.R. Acad. Sci., Paris*, **229**, 27.   [*53*]

GRIVET, P., SOUTIF, M. and GABILLARD, R. (1951). *Physica*, **17**, 420.   [*48*]

GUPTILL, E. W., ARCHIBALD, W. J. and WARREN, E. S. (1950). *Canad. J. Res. A*, **28**, 359.   [*228*]

GUTOWSKY, H. S. (1951). *Phys. Rev.* **83**, 1073.   [*199, 200, 241*]

GUTOWSKY, H. S. and HOFFMAN, C. J. (1950). *Phys. Rev.* **80**, 110.   [*142*]

GUTOWSKY, H. S. and HOFFMAN, C. J. (1951). *J. Chem. Phys.* **19**, 1259.   [*73, 142*]

GUTOWSKY, H. S., KISTIAKOWSKY, G. B., PAKE, G. E. and PURCELL,
E. M. (1949). *J. Chem. Phys.* **17**, 972.   [*156, 158, 239*]
GUTOWSKY, H. S. and McCALL, D. W. (1951). *Phys. Rev.* **82**, 748.   [*144*]
GUTOWSKY, H. S. and McCALL, D. W. (1954). *J. Chem. Phys.* **22**,
162.   [*142*]
GUTOWSKY, H. S., McCALL, D. W., McGARVEY, B. R. and MEYER,
L. H. (1951). *J. Chem. Phys.* **19**, 1328.   [*142*]
GUTOWSKY, H. S., McCALL, D. W., McGARVEY, B. R. and MEYER,
L. H. (1952). *J. Amer. Chem. Soc.* **74**, 4809.   [*142*]
GUTOWSKY, H. S., McCALL, D. W. and SLICHTER, C. P. (1951).
Phys. Rev. **84**, 589.   [*144, 145*]
GUTOWSKY, H. S., McCALL, D. W. and SLICHTER, C. P. (1953).
J. Chem. Phys. **21**, 279.   [*144*]
GUTOWSKY, H. S. and McCLURE, R. E. (1951). *Phys. Rev.* **81**, 276.   [*80*]
GUTOWSKY, H. S., McCLURE, R. E. and HOFFMAN, C. J. (1951).
Phys. Rev. **81**, 635.   [*86, 228, 239*]
GUTOWSKY, H. S. and McGARVEY, B. R. (1952). *J. Chem. Phys.*
20, 1472.   [*200, 201, 202, 241*]
GUTOWSKY, H. S. and McGARVEY, B. R. (1953*a*). *Phys. Rev.* **91**, 81.
                                              [*75, 142, 228*]
GUTOWSKY, H. S. and McGARVEY, B. R. (1953*b*). *J. Chem. Phys.*
21, 1423.   [*142, 239*]
GUTOWSKY, H. S. and MEYER, L. H. (1953). *J. Chem. Phys.* **21**,
2122.   [*240*]
GUTOWSKY, H. S., MEYER, L. H. and McCLURE, R. E. (1953).
Rev. Sci. Instrum. **24**, 644.   [*53, 55, 73*]
GUTOWSKY, H. S. and PAKE, G. E. (1948). *J. Chem. Phys.* **16**, 1164.   [*239*]
GUTOWSKY, H. S. and PAKE, G. E. (1950). *J. Chem. Phys.* **18**, 162.
                                          [*165–170, 173, 239*]
GUTOWSKY, H. S., PAKE, G. E. and BERSOHN, R. (1954). *J. Chem.
Phys.* **22**, 643.   [*162,164, 239*]
GUTOWSKY, H. S. and SAIKA, A. (1953). *J. Chem. Phys.* **21**, 1688.   [*142*]
GVOZDOVER, S. D. and IEVSKAYA, N. M. (1953). *J. Exp. Theor. Phys.*
25, 435.   [*59*]
GVOZDOVER, S. D. and MAGAZANIK, A. A. (1950). *J. Exp. Theor.
Phys.* **20**, 705.   [*132*]
HAHN, E. L. (1949). *Phys. Rev.* **76**, 145.   [*138*]
HAHN, E. L. (1950*a*). *Phys. Rev.* **77**, 297.   [*139*]
HAHN, E. L. (1950*b*). *Phys. Rev.* **80**, 580.   [*62, 63, 127, 136, 139*]
HAHN, E. L. (1953). *Phys. To-day*, **6**, no. 11, p. 4.   [*140*]
HAHN, E. L. and MAXWELL, D. E. (1951). *Phys. Rev.* **84**, 1246.   [*144, 145*]
HAHN, E. L. and MAXWELL, D. E. (1952). *Phys. Rev.* **88**, 1070.   [*144*]
HATTON, J. and ROLLIN, B. V. (1949). *Proc. Roy. Soc.* A, **199**, 222.
                              [*174, 176, 182, 195, 239, 241*]
HATTON, J., ROLLIN, B. V. and SEYMOUR, E. F. W. (1951). *Phys.
Rev.* **83**, 672.   [*228, 237*]
HAWKINS, W. B. and DICKE, R. H. (1953). *Phys. Rev.* **91**, 1008.   [*65*]
HEDGRAN, A. (1951). *Phys. Rev.* **82**, 128.   [*104*]
HEITLER, W. and TELLER, E. (1936). *Proc. Roy. Soc.* A, **155**, 629.   [*193*]

HICKMOTT, T. W. and SELWOOD, P. W. (1952). *J. Chem. Phys.* **20**, 1339. [*147*]

HIPPLE, J. A., SOMMER, H. and THOMAS, H. A. (1949). *Phys. Rev.* **76**, 1877. [*91, 227*]

HOLROYD, L. V., CODRINGTON, R. S., MROWCA, B. A. and GUTH, E. (1951). *J. Appl. Phys.* **22**, 696. [*240*]

HOPKINS, N. J. (1949). *Rev. Sci. Instrum.* **20**, 401. [*51, 98*]

HUNTEN, D. M. (1950). *Phys. Rev.* **78**, 806. [*228*]

HUNTEN, D. M. (1951). *Canad. J. Phys.* **29**, 463. [*228*]

HYLLERAAS, E. and SKAVLEM, S. (1950). *Phys. Rev.* **79**, 117. [*80*]

ITOH, J., KUSAKA, R., KIRIYAMA, R. and YABUMOTO, S. (1953). *J. Chem. Phys.* **21**, 1895. [*240*]

ITOH, J., KUSAKA, R. and YAMAGATA, Y. (1954). *J. Phys. Soc. Japan,* **9**, 209. [*240*]

ITOH, J., KUSAKA, R., YAMAGATA, Y., KIRIYAMA, R. and IBAMOTO, H. (1952). *J. Chem. Phys.* **20**, 1503; **21**, 190. [*158, 239*]

ITOH, J., KUSAKA, R., YAMAGATA, Y., KIRIYAMA, R. and IBAMOTO, H. (1953a). *Physica,* **19**, 415. [*239*]

ITOH, J., KUSAKA, R., YAMAGATA, Y., KIRIYAMA, R. and IBAMOTO, H. (1953b). *J. Phys. Soc. Japan,* **8**, 293. [*158, 239*]

ITOH, J., KUSAKA, R., YAMAGATA, Y., KIRIYAMA, R., IBAMOTO, H., KANDA, T. and MASUDA, Y. (1953). *J. Phys. Soc. Japan,* **8**, 287. [*239*]

JACCARINO, V., KING, J. G., SATTEN, R. A. and STROKE, H. H. (1954). *Phys. Rev.* **94**, 1798. [*88*]

JACOBSOHN, B., ANDERSON, W. A. and ARNOLD, J. T. (1954). *Nature, Lond.* **173**, 772. [*146*]

JACOBSOHN, B. A. and WANGSNESS, R. K. (1948). *Phys. Rev.* **73**, 942. [*133*]

JARRETT, H. S., SADLER, M. S. and SHOOLERY, J. N. (1953). *J. Chem. Phys.* **21**, 2092. [*143*]

JEFFRIES, C. D. (1951). *Phys. Rev.* **81**, 1040. [*91, 92, 227*]

JEFFRIES, C. D. (1953a). *Phys. Rev.* **90**, 1130. [*228*]

JEFFRIES, C. D. (1953b). *Phys. Rev.* **92**, 1262. [*228*]

JEFFRIES, C. D., LÖLIGER, H. and STAUB, H. H. (1951). *Helv. phys. acta,* **24**, 643. [*228*]

JEFFRIES, C. D., LÖLIGER, H. and STAUB, H. H. (1952). *Phys. Rev.* **85**, 478. [*228*]

JEFFRIES, C. D. and SOGO, P. B. (1953). *Phys. Rev.* **91**, 1286. [*228*]

JONES, H. and SCHIFF, B. (1954). *Proc. Phys. Soc. Lond.* **67**, 217. [*197, 198*]

KAKIUCHI, Y. and KOMATSU, H. (1952). *J. Phys. Soc. Japan,* **7**, 380. [*239*]

KAKIUCHI, Y., SHONO, H., KOMATSU, H. and KIGOSHI, K. (1951). *J. Chem. Phys.* **19**, 1069. [*158, 239*]

KAKIUCHI, Y., SHONO, H., KOMATSU, H. and KIGOSHI, K. (1952). *J. Phys. Soc. Japan,* **7**, 102. [*53, 158, 239*]

KAMBE, K. and USUI, T. (1952). *Progr. Theor. Phys.* **8**, 302. [*159*]

KANDA, T. (1952). *J. Phys. Soc. Japan,* **7**, 296. [*228*]

KANDA, T., MASUDA, Y., KUSAKA, R., YAMAGATA, Y. and ITOH, J. (1951). *Phys. Rev.* **83**, 1066. [*228*]

KANDA, T., MASUDA, Y., KUSAKA, R., YAMAGATA, Y. and ITOH, J. (1952). *Phys. Rev.* **85**, 938. [*228*]

KARPLUS, R. and KROLL, N. (1950). *Phys. Rev.* **77**, 536. [*95*]

KASTLER, A. (1950). *J. Phys. Radium*, 11, 255.   [65]

KELLOGG, J. M. B., RABI, I. I., RAMSEY, N. F. and ZACHARIAS,
   J. R. (1939). *Phys. Rev.* 56, 728.   [82]

KHUTSISHVILI, G. R. (1952). *J. Exp. Theor. Phys.* 22, 382.   [176]

KNIGHT, W. D. (1949). *Phys. Rev.* 76, 1259.   [141, 196, 202, 241]

KNIGHT, W. D. (1952). *Phys. Rev.* 85, 762.   [241]

KNIGHT, W. D. (1953). *Phys. Rev.* 92, 539.   [201, 241]

KNIGHT, W. D. and COHEN, V. W. (1949). *Phys. Rev.* 76, 1421. [228, 239]

KNIGHT, W. D. and KITTEL, C. (1952). *Phys. Rev.* 86, 573.   [190]

KNOEBEL, H. W. and HAHN, E. L. (1951). *Rev. Sci. Instrum.* 22, 904.
                                                          [53, 98, 101]

KOENIG, S. H., PRODELL, A. G. and KUSCH, P. (1952). *Phys. Rev.*
   88, 191.   [95]

KOHN, W. and BLOEMBERGEN, N. (1950). *Phys. Rev.* 80, 913; 82, 283. [197]

KOJIMA, S. and OGAWA, S. (1953). *J. Phys. Soc. Japan*, 8, 283.   [240]

KOJIMA, S., OGAWA, S. and TORIZUKA, K. (1951). *Sci. of Light*,
   Tokyo, 1, 101.   [55, 240]

KORRINGA, J. (1950). *Physica*, 16, 601.   [194, 195, 198]

KRAMERS, H. A. (1927). *Atti del Congresso Internationale dei Fisici*,
   Como, 2, 545.   [23]

KRONIG, R. DE L. (1926). *J. Opt. Soc. Amer.* 12, 547.   [23]

KRÜGER, H. (1951). *Z. Phys.* 130, 371.   [220]

KUBO, R. and TOMITA, K. (1954). *J. Phys. Soc. Japan*, 9, 888. [150, 234]

KUSCH, P. and ECK, T. G. (1954). *Phys. Rev.* 94, 1799.   [88]

LAMARCHE, G. and VOLKOFF, G. M. (1953). *Canad. J. Phys.* 31, 1010.[220]

LAMB, W. E. (1941). *Phys. Rev.* 60, 817.   [79]

LANGEVIN, P. (1905). *J. Phys. théor. appl.* (4), 4, 678.   [22]

LASAREW, B. G. and SCHUBNIKOW, L. W. (1937). *Phys. Z. Sowjet.*
   11, 445.   [2, 22]

LEVINTHAL, E. C. (1950). *Phys. Rev.* 78, 204.   [59, 227]

LEVY, H. A. and PETERSON, S. W. (1952). *Phys. Rev.* 86, 766.   [162]

LIDDEL, U. and RAMSEY, N. F. (1951). *J. Chem. Phys.* 19, 1608.   [142]

LINDSTRÖM, G. (1950). *Phys. Rev.* 78, 817.   [227]

LINDSTRÖM, G. (1951a). *Physica*, 17, 412.   [227]

LINDSTRÖM, G. (1951b). *Ark. Fys.* 4, 1.   [101, 102, 227, 228]

LINDSTRÖM, G. (1951c). *Phys. Rev.* 83, 465.   [102, 227]

LINDSTRÖM, G. (1952). *Phys. Rev.* 87, 678.   [102]

LORENTZ, H. A. (1909). *The Theory of Electrons*, p. 138. Reprinted
   1952. New York: Dover.   [78]

McGARVEY, B. R. and GUTOWSKY, H. S. (1953). *J. Chem. Phys.* 21,
   2114.   [199, 201, 202, 241]

McNEIL, E. B., SLICHTER, C. P. and GUTOWSKY, H. S. (1951).
   Phys. Rev. 84, 1245.   [144]

MANUS, C., MERCIER, R., DENIS, P., BÉNÉ, G. and EXTERMANN, R.
   (1954). *C.R. Acad. Sci., Paris*, 238, 1315.   [136]

MARGENAU, H. and MURPHY, G. M. (1943). *The Mathematics of
   Physics and Chemistry*. New York: D. van Nostrand Com-
   pany Inc.   [168]

MASUDA, Y. and KANDA, T. (1953). *J. Phys. Soc. Japan*, 8, 432.   [142]

MASUDA, Y. and KANDA, T. (1954). *J. Phys. Soc. Japan*, **9**, 82.   [*142*]

MEYER, L. H. and GUTOWSKY, H. S. (1953). *J. Phys. Chem.* **57**, 481. [*142*]

MEYER, L. H., SAIKA, A. and GUTOWSKY, H. S. (1953). *J. Amer. Chem. Soc.* **75**, 4567.   [*142*]

MOOI, J. and SELWOOD, P. W. (1952). *J. Amer. Chem. Soc.* **74**, 2461 [*147*]

MOTT, N. F. and JONES, H. (1936). *The Theory of the Properties of Metals and Alloys.* Oxford University Press.   [*198*]

MOXON, L. A. (1949). *Recent Advances in Radio Receivers.* Cambridge University Press .   [*68*]

MUTO, T. and WATANABE, M. (1952). *Progr. Theor. Phys.* **8**, 231.   [*176*]

NACHTRIEB, N. H., CATALANO, E. and WEIL, J. A. (1952). *J. Chem. Phys.* **20**, 1185.   [*201*]

NEWELL, G. F. (1950). *Phys. Rev.* **80**, 476.   [*81*]

NEWMAN, R. (1950*a*). *J. Chem. Phys.* **18**, 669.   [*166, 184, 239*]

NEWMAN, R. (1950*b*). *J. Chem. Phys.* **18**, 1303.   [*240*]

NEWMAN, R. and OGG, R. A. (1951). *J. Chem. Phys.* **19**, 214.   [*147*]

NORBERG, R. E. (1952). *Phys. Rev.* **86**, 745.   [*149*]

NORBERG, R. E. and SLICHTER, C. P. (1951). *Phys. Rev.* **83**, 1074
[*195, 201, 241*]

OGG, R. A. (1954). *J. Chem. Phys.* **22**, 560.   [*142*]

OGG, R. A. and RAY, J. D. (1954). *J. Chem. Phys.* **22**, 147.   [*86, 228*]

OVERHAUSER, A. W. (1953). *Phys. Rev.* **92**, 411.   [*195*]

PACKARD, M. E. (1948). *Rev. Sci. Instrum.* **19**, 435.   [*59, 101*]

PACKARD, M. E. and VARIAN, R. (1954). *Phys. Rev.* **93**, 941.   [*97*]

PACKARD, M. E. and WEAVER, H. E. (1952). *Phys. Rev.* **88**, 163.   [*148*]

PAKE, G. E. (1948). *J. Chem. Phys.* **16**, 327.
[*46, 47, 73, 152-4, 161, 239, 242*]

PAKE, G. E. (1953). *J. Chim. phys.* **50**, C 104.   [*145*]

PAKE, G. E. and GUTOWSKY, H. S. (1948). *Phys. Rev.* **74**, 979.   [*239*]

PAKE, G. E. and PURCELL, E. M. (1948). *Phys. Rev.* **74**, 1184; **75**,.
534 .   [*24, 30, 161, 239*]

PAULI, W. (1924). *Naturwissenschaften*, **12**, 741.   [*1*]

PEKÁREK, L. and URBANEC, J. (1952). *Czech. J. Phys.* **1**, 78.   [*53*]

PERLMAN, M. M. and BLOOM, M. (1953). *Phys. Rev.* **88**, 1290.   [*164*]

PETCH, H. E., CRANNA, N. G. and VOLKOFF, G. M. (1953). *Canad. J. Phys.* **31**, 837.   [*210, 228, 239*]

PETCH, H. E., SMELLIE, D. W. and VOLKOFF, G. M. (1951). *Phys. Rev.* **84**, 602.   [*210, 239*]

PETCH, H. E., VOLKOFF, G. M. and CRANNA, N. G. (1952). *Phys. Rev.* **88**, 1201.   [*210, 239*]

PORTIS, A. M. (1953). *Phys. Rev.* **91**, 1071.   [*2*]

POSS, H. L. (1947). *Phys. Rev.* **72**, 637.   [*228*]

POSS, H. L. (1949). *Phys. Rev.* **75**, 600.   [*228*]

POULIS, N. J. (1950). *Physica*, **16**, 373.   [*195, 241*]

POULIS, N. J. (1951). *Physica*, **17**, 392.   [*53, 185, 187, 239*]

POULIS, N. J., VAN DEN HANDEL, J., UBBINK, J., POULIS, J. A. and GORTER, C. J. (1951). *Phys. Rev.* **82**, 552.   [*239*]

POULIS, N. J. and HARDEMAN, G. E. G. (1952*a*). *Physica*, **18**, 201.
[*187-190, 239*]

POULIS, N. J. and HARDEMAN, G. E. G. (1952b). *Physica*, **18**, 315.
[*187, 189, 239*]
POULIS, N. J. and HARDEMAN, G. E. G. (1953a). *Physica*, **19**, 391. [*189, 239*]
POULIS, N. J. and HARDEMAN, G. E. G. (1953b). *J. Chim. phys.* **50**,
C110. [*239*]
POULIS, N. J. and HARDEMAN, G. E. G. (1954). *Physica*, **20**, 7. [*189, 239*]
POULIS, N. J., HARDEMAN, G. E. G. and BÖLGER, B. (1952).
Physica, **18**, 429. [*189, 239*]
POUND, R. V. (1947a). *Phys. Rev.* **72**, 527. [*51*]
POUND, R. V. (1947b). *Phys. Rev.* **72**, 1273. [*87, 205, 228*]
POUND, R. V. (1948a). *Phys. Rev.* **73**, 523. [*228, 239*]
POUND, R. V. (1948b). *Phys. Rev.* **73**, 1112. [*87, 191, 228, 239, 241*]
POUND, R. V. (1949). *Phys. Rev.* **76**, 1410. [*220*]
POUND, R. V. (1950). *Phys. Rev.* **79**, 685. [*73, 88, 206–212, 228, 239*]
POUND, R. V. (1951). *Phys. Rev.* **81**, 156. [*179, 239*]
POUND, R. V. (1952). *Progr. Nucl. Phys.* **2**, 21. [*51, 85*]
POUND, R. V. (1953). *J. Phys. Chem.* **57**, 743. [*211*]
POUND, R. V. and KNIGHT, W. D. (1950). *Rev. Sci. Instrum.* **21**, 219.
[*51, 98, 101, 146*]
POWLES, J. G. and GUTOWSKY, H. S. (1953a). *J. Chem. Phys.* **21**,
1695. [*171, 175, 239*]
POWLES, J. G. and GUTOWSKY, H. S. (1953b). *J. Chem. Phys.* **21**, 1704. [*240*]
PRATT, L. and RICHARDS, R. E. (1953). *Trans. Faraday Soc.* **49**, 744. [*239*]
PRATT, L. and RICHARDS, R. E. (1954). *Trans. Faraday Soc.* **50**, 670. [*240*]
PROCTOR, W. G. (1949a). *Phys. Rev.* **75**, 522. [*228*]
PROCTOR, W. G. (1949b). *Phys. Rev.* **76**, 684. [*228*]
PROCTOR, W. G. (1950). *Phys. Rev.* **79**, 35. [*60, 228*]
PROCTOR, W. G. and YU, F. C. (1949). *Phys. Rev.* **76**, 1728. [*228*]
PROCTOR, W. G. and YU, F. C. (1950a). *Phys. Rev.* **77**, 716. [*228*]
PROCTOR, W. G. and YU, F. C. (1950b). *Phys. Rev.* **78**, 471. [*143, 228*]
PROCTOR, W. G. and YU, F. C. (1950c). *Phys. Rev.* **77**, 717. [*141, 142*]
PROCTOR, W. G. and YU, F. C. (1951). *Phys. Rev.* **81**, 20.
[*76, 80, 88, 141–3, 149, 228*]
PRODELL, A. G. and KUSCH, P. (1952). *Phys. Rev.* **88**, 184. [*96*]
PRYCE, M. H. L. and STEVENS, K. W. H. (1950). *Proc. Phys. Soc.*
A **63**, 36. [*159*]
PURCELL, E. M. (1946). *Phys. Rev.* **69**, 681. [*12*]
PURCELL, E. M. (1951). *Physica*, **17**, 282. [*176*]
PURCELL, E. M. (1953). Private communication. [*125*]
PURCELL, E. M., BLOEMBERGEN, N. and POUND, R. V. (1946).
Phys. Rev. **70**, 988. [*239*]
PURCELL, E. M. and POUND, R. V. (1951). *Phys. Rev.* **81**, 279. [*180, 181, 239*]
PURCELL, E. M., POUND, R. V. and BLOEMBERGEN, N. (1946).
Phys. Rev. **70**, 986. [*147, 148*]
PURCELL, E. M., TORREY, H. C. and POUND, R. V. (1946). *Phys.
Rev.* **69**, 37. [*5, 40, 239*]
QUINN, W. E. and BROWN, R. M. (1953). *J. Chem. Phys.* **21**, 1605. [*144*]
RABI, I. I., MILLMAN, S., KUSCH, P. and ZACHARIAS, J. R. (1939).
Phys. Rev. **55**, 526. [*3*]

RABI, I. I., RAMSEY, N. F. and SCHWINGER, J. (1954). *Rev. Mod. Phys.* **26**, 167.   [*137*]

RAMSEY, N. F. (1950*a*). *Phys. Rev.* **77**, 567.   [*80, 142*]

RAMSEY, N. F. (1950*b*). *Phys. Rev.* **78**, 699.   [*80, 90, 142*]

RAMSEY, N. F. (1950*c*). *Phys. Rev.* **79**, 1010.   [*228*]

RAMSEY, N. F. (1951). *Physica*, **17**, 303.   [*80, 142*]

RAMSEY, N. F. (1952*a*). *Phys. Rev.* **85**, 688.   [*81*]

RAMSEY, N. F. (1952*b*). *Phys. Rev.* **86**, 243.   [*80, 142*]

RAMSEY, N. F. (1953). *Phys. Rev.* **91**, 303.   [*145*]

RAMSEY, N. F. and POUND, R. V. (1951). *Phys. Rev.* **81**, 278.   [*181, 239*]

RAMSEY, N. F. and PURCELL, E. M. (1952). *Phys. Rev.* **85**, 143.   [*145*]

REIF, F. and PURCELL, E. M. (1953). *Phys. Rev.* **91**, 631.   [*182–4, 239*]

RICHARDS, R. E. and SMITH, J. A. S. (1951). *Trans. Faraday Soc.* **47**, 1261.   [*157, 158, 239*]

RICHTER, H. L., HUMPHREY, F. B. and YOST, D. M. (1954). *Rev. Sci. Instrum.* **25**, 190.   [*97*]

ROBERTS, A. (1947*a*). *Rev. Sci. Instrum.* **18**, 845.   [*53, 55*]

ROBERTS, A. (1947*b*). *Phys. Rev.* **72**, 979.   [*227*]

ROGERS, E. H. and STAUB, H. H. (1949). *Phys. Rev.* **76**, 980.   [*82, 227*]

ROLLIN, B. V. (1946). *Nature, Lond.* **158**, 669.   [*35, 239*]

ROLLIN, B. V. (1949). *Rep. Progr. Phys.* **12**, 22.   [*35*]

ROLLIN, B. V. and HATTON, J. (1947). *Nature, Lond.* **159**, 201.   [*239*]

ROLLIN, B. V. and HATTON, J. (1948). *Phys. Rev.* **74**, 346.
    [*176, 194, 239, 241*]

ROLLIN, B. V., HATTON, J., COOKE, A. H. and BENZIE, R. J. (1947). *Nature, Lond.* **160**, 436.   [*239*]

ROSE, M. E. (1938). *Phys. Rev.* **53**, 715.   [*73*]

ROSS, I. M. and JOHNSON, F. B. (1951). *Nature, Lond.* **167**, 286.   [*46*]

RUDERMAN, M. A. and KITTEL, C. (1954). *Phys. Rev.* **96**, 99.   [*199*]

RUSHWORTH, F. A. (1952). *J. Chem. Phys.* **20**, 920.   [*173, 240*]

RUSHWORTH, F. A. (1954). *Proc. Roy. Soc.* A, **222**, 526.   [*175, 239*]

SACHS, A. M. and TURNER, E. (1951). Quoted by PURCELL (1951).
    [*20, 177, 239*]

SAIKA, A. and SLICHTER, C. P. (1954). *J. Chem. Phys.* **22**, 26.   [*142*]

SALPETER, E. E. (1950). *Proc. Phys. Soc. Lond.* A, **63**, 337.   [*129*]

SANDS, R. H. and PAKE, G. E. (1953). *Phys. Rev.* **89**, 896.   [*85*]

SAXTON, J. A. (1946). *Meteorological Factors in Radio-wave Propagation Conference*, pp. 278–305. *Phys. Soc., Lond.*   [*116*]

SCHNEIDER, E. E. (1948). *Proc. Phys. Soc.* **61**, 569.   [*46*]

SCHUSTER, N. A. (1951). *Rev. Sci. Instrum.* **22**, 254.   [*45*]

SCHUSTER, N. A. and PAKE, G. E. (1951*a*). *Phys. Rev.* **81**, 157. [*87, 228, 239*]

SCHUSTER, N. A. and PAKE, G. E. (1951*b*). *Phys. Rev.* **81**, 886.   [*228, 239*]

SCHWINGER, J. (1948). *Phys. Rev.* **73**, 416.   [*95*]

SEITZ, F. (1951). *Phase Transformations in Solids*, ed. Smoluchowski *et al.*, p. 77. New York: John Wiley and Sons, Inc.   [*174*]

SELWOOD, P. W. and SCHROYER, F. K. (1950). *Disc. Faraday Soc.* **8**, 337.   [*147*]

SEYMOUR, E. F. W. (1953). *Proc. Phys. Soc. Lond.* A, **66**, 85.   [*201, 241*]

SHAW, T. M. and ELSKEN, R. H. (1950). *J. Chem. Phys.* **18**, 1113.   [*146*]

SHAW, T. M. and ELSKEN, R. H. (1953). *J. Chem. Phys.* **21**, 565. [*146*]
SHAW, T. M., ELSKEN, R. H. and KUNSMAN, C. H. (1953). *J. Assoc. Offic. Agric. Chem.* **36**, 1070. [*146*]
SHAW, T. M., ELSKEN, R. H. and PALMER, K. J. (1952). *Phys. Rev.* **85**, 762. [*240*]
SHERIFF, R. E., CHAMBERS, W. H. and WILLIAMS, D. (1950). *Phys. Rev.* **78**, 476. [*228*]
SHERIFF, R. E. and WILLIAMS, D. (1950). *Phys. Rev.* **79**, 175. [*228*]
SHERIFF, R. E. and WILLIAMS, D. (1951). *Phys. Rev.* **82**, 651. [*228*]
SHERMAN, C. (1954). *Phys. Rev.* **93**, 1429. [*112*]
SHOOLERY, J. N. (1953). *J. Chem. Phys.* **21**, 1899. [*142, 144*]
SIEGBAHN, K. (1944). *Ark. Mat. Astr. Fys.* **30**A, no. 20. [*103*]
SIEGBAHN, K. and LINDSTRÖM, G. (1949*a*). *Nature, Lond.* **163**, 211. [*227*]
SIEGBAHN, K. and LINDSTRÖM, G. (1949*b*). *Ark. Fys.* **1**, 193. [*227*]
SLATER, J. C. (1941). *J. Chem. Phys.* **9**, 16. [*185*]
SMALLER, B. (1951). *Phys. Rev.* **83**, 812. [*81, 146, 227*]
SMALLER, B., YASAITIS, E. and ANDERSON, H. L. (1951). *Phys. Rev.* **81**, 896. [*227*]
SMALLER, B., YASAITIS, E., AVERY, E. C. and HUTCHISON, D. A. (1952). *Phys. Rev.* **88**, 414. [*145*]
SMITH, J. A. S. and RICHARDS, R. E. (1952). *Trans. Faraday Soc.* **48**, 307. [*239*]
SOGO, P. B. and JEFFRIES, C. D. (1954). *Phys. Rev.* **93**, 174. [*75, 228, 241*]
SOMMER, H., THOMAS, H. A. and HIPPLE, J. A. (1950). *Phys. Rev.* **80**, 487. [*91, 227*]
SOMMER, H., THOMAS, H. A. and HIPPLE, J. A. (1951). *Phys. Rev.* **82**, 697. [*91, 227*]
SOUTIF, M. (1951). *Rev. Sci., Paris*, **89**, 203. [*48, 53, 55, 136, 240*]
SOUTIF, M. and AYANT, Y. (1953). *J. Chim. phys.* **50**, C 107. [*239*]
SOUTIF, M., DREYFUS, B. and AYANT, Y. (1951). *C.R. Acad. Sci., Paris*, **233**, 395. [*239*]
SOUTIF, M. and GABILLARD, R. (1950). *C.R. Acad. Sci., Paris*, **230**, 2012. [*53*]
SOUTIF, M. and GABILLARD, R. (1951). *Physica*, **17**, 319. [*136*]
SPENCE, R. D., GUTOWSKY, H. S. and HOLM, C. H. (1953). *J. Chem. Phys.* **21**, 1891. [*240*]
SPENCE, R. D., MOSES, H. A. and JAIN, P. L. (1953). *J. Chem. Phys.* **21**, 380. [*240*]
SPOONER, R. B. and SELWOOD, P. W. (1949). *J. Amer. Chem. Soc.* **71**, 2184. [*147*]
STAUB, H. H. and ROGERS, E. H. (1950). *Helv. phys. acta*, **23**, 63. [*82, 227*]
STERN, O. (1921). *Z. Phys.* **7**, 249. [*2*]
STRICK, E., BRADFORD, R., CLAY, C. and CRAFT, A. (1951). *Phys. Rev.* **84**, 363. [*141*]
STUECKELBERG, E. C. G. (1948). *Phys. Rev.* **73**, 808. [*100*]
SURYAN, G. (1950). *Phys. Rev.* **80**, 119. [*46*]
SURYAN, G. (1951). *Proc. Indian Acad. Sci.* A, **33**, 107. [*112*]
SURYAN, G. (1953). *J. Indian Inst. Sci.* A, **35**, 25. [*55*]
TAYLOR, K. (1953). *Nature, Lond.* **172**, 722. [*136, 239*]

THOMAS, H. A. (1950). *Phys. Rev.* **80**, 901. [*80, 90*]

THOMAS, H. A. (1952). *Electronics*, **25**, 114 [*101*]

THOMAS, H. A., DRISCOLL, R. L. and HIPPLE, J. A. (1949). *Phys. Rev.* **75**, 902. [*88, 104, 227*]

THOMAS, H. A., DRISCOLL, R. L. and HIPPLE, J. A. (1950 *a*). *Phys. Rev.* **78**, 787. [*88, 104, 227*]

THOMAS, H. A., DRISCOLL, R. L. and HIPPLE, J. A. (1950 *b*). *J. Res. Nat. Bur. Stand.* **44**, 569. [*88, 101, 104, 227*]

THOMAS, H. A. and HUNTOON, R. D. (1949). *Rev. Sci. Instrum.* **20**, 516. [*47, 48, 89*]

THOMAS, J. T., ALPERT, N. L. and TORREY, H. C. (1950). *J. Chem. Phys.* **18**, 1511. [*175, 239*]

TING, Y. and WILLIAMS, D. (1953). *Phys. Rev.* **89**, 595. [*228*]

TOMITA, K. (1952). *Progr. Theor. Phys.* **8**, 138. [*158, 239*]

TOMITA, K. (1953). *Phys. Rev.* **89**, 429. [*158, 239*]

TOMITA, K. and MANNARI, I. (1953). *Progr. Theor. Phys.* **10**, 367. [*240*]

TORREY, H. C. (1949). *Phys. Rev.* **76**, 1059. [*46, 47, 62, 63, 71, 136, 137*]

TORREY, H. C. (1952). *Phys. Rev.* **85**, 365. [*141*]

TORREY, H. C. (1953). *Phys. Rev.* **92**, 962. [*117*]

TOWNES, C. H., HERRING, C. and KNIGHT, W. D. (1950). *Phys. Rev.* **77**, 852. [*196, 197, 241*]

TUTTLE, W. N. (1940). *Proc. Inst. Radio Engrs., N.Y.* **28**, 23. [*48*]

VAN VLECK, J. H. (1948). *Phys. Rev.* **74**, 1168. [*158, 161, 165, 240*]

VERBRUGGE, F. and HENRY, R. L. (1951). *Phys. Rev.* **83**, 211. [*148*]

VOLKOFF, G. M. (1953). *Canad. J. Phys.* **31**, 820. [*210*]

VOLKOFF, G. M., PETCH, H. E. and SMELLIE, D. W. L. (1952). *Canad. J. Phys.* **30**, 270. [*53, 210, 239*]

WALCHLI, H. E. (1953 *a*). *Phys. Rev.* **90**, 331. [*228*]

WALCHLI, H. E. (1953 *b*). *U.S. Atomic Energy Commission Report ORNL*—1469. *Supplement* (1955) [*224*]

WALCHLI, H. E., LEYSHON, W. E. and SCHEITLIN, F. M. (1952). *Phys. Rev.* **85**, 922. [*228*]

WALCHLI, H. E., LIVINGSTON, R. and HEBERT, G. (1951). *Phys. Rev.* **82**, 97. [*228*]

WALCHLI, H. E., LIVINGSTON, R. and MARTIN, W. J. (1952). *Phys. Rev.* **85**, 479. [*228*]

WALCHLI, H. E. and MORGAN, H. W. (1952). *Phys. Rev.* **87**, 541. [*229, 241*]

WALLER, I. (1932). *Z. Phys.* **79**, 370. [*175, 211*]

WANG, M. C. and UHLENBECK, G. E. (1945). *Rev. Mod. Phys.* **17**, 323. [*232*]

WANGSNESS, R. K. and BLOCH, F. (1953). *Phys. Rev.* **89**, 728. [*26*]

WARING, C. E., SPENCER, R. H. and CUSTER, R. L. (1952). *Rev. Sci. Instrum.* **23**, 497. [*47, 48*]

WATKINS, G. D. and POUND, R. V. (1951). *Phys. Rev.* **82**, 343. [*51, 228*]

WATKINS, G. D. and POUND, R. V. (1953). *Phys. Rev.* **89**, 658 [*85, 210, 239*]

WAUGH, J. S., HUMPHREY, F. B. and YOST, D. M. (1953). *J. Phys. Chem.* **57**, 486. [*158, 240*]

WEAVER, H. E. (1953). *Phys. Rev.* **89**, 923. [*60, 228, 239*]

WHITE, H. E. (1934). *Introduction to Atomic Spectra.* McGraw-Hill. [*30*]

WHITEHEAD, J. R. (1950). *Super-regenerative Receivers.* Cambridge University Press. [*54*]

WILLIAMS, D. (1951). *Physica,* **17**, 454. [*55*]

WILLIAMS, G. A., McCALL, D. W. and GUTOWSKY, H. S. (1954). *Phys. Rev.* **93**, 1428. [*86, 229*]

WILSON, G. W. and PAKE, G. E. (1953). *J. Polymer Sci.* **10**, 503. [*240*]

WIMETT, T. F. (1953*a*). *Phys. Rev.* **91**, 476. [*145*]

WIMETT, T. F. (1953*b*). *Phys. Rev.* **91**, 499. [*73, 81, 228*]

YASAITIS, E. and SMALLER, B. (1951). *Phys. Rev.* **82**, 750. [*228*]

YOKOTA, I. (1952). *Progr. Theor. Phys.* **8**, 380. [*159*]

ZENER, C. (1951). *Phys. Rev.* **81**, 440. [*190*]

ZIMMERMAN, J. R. (1953). *J. Chem. Phys.* **21**, 1605. [*128*]

ZIMMERMAN, J. R. (1954). *J. Chem. Phys.* **22**, 950. [*128*]

ZIMMERMAN, J. R. and WILLIAMS, D. (1948). *Phys. Rev.* **74**, 1885. [*228*]

ZIMMERMAN, J. R. and WILLIAMS, D. (1949*a*). *Phys. Rev.* **75**, 198. [*228*]

ZIMMERMAN, J. R. and WILLIAMS, D. (1949*b*). *Phys. Rev.* **75**, 699. [*228*]

ZIMMERMAN, J. R. and WILLIAMS, D. (1949*c*). *Phys. Rev.* **76**, 350. [*227*]

SUBJECT INDEX